SLATE ROOFING IN CANADA

Mary Cullen

Studies in Archaeology,
Architecture and History

National Historic Parks and Sites
Parks Service
Environment Canada

Available in Canada through authorized bookstore agents and other bookstores, or by mail from the Canadian Government Publishing Centre, Supply and Services Canada, Hull, Quebec, Canada K1A 0S9.

En français ce numéro s'intitule: *Les couvertures en ardoises au Canada* (no de catalogue R61-2/9-46F). En vente au Canada par l'entremise de nos agents libraires agréés et autres librairies, ou par la poste au Centre d'édition du gouvernement du Canada, Approvisionnements et Services Canada, Hull, Québec, Canada K1A 0S9.

Published under the authority
of the Minister of the Environment,
Ottawa, 1990.

Editing and design: Sheila Ascroft
Cover design: Rod Won
Front and back cover photos: The Keeper's House at the Necropolis, Toronto, Ontario.

Parks publishes the results of its research in archaeology, architecture and history. A list of titles is available from Research Publications, Parks Service, Environment Canada, 1600 Liverpool Court, Ottawa, Ontario K1A 0H3.

Canadian Cataloguing in Publication Data

Cullen, Mary

Slate roofing in Canada

(Studies in archaeology, architecture and history,
 ISSN 0821-1027)
Issued also in French under title: Les couvertures en ardoises au Canada.
Includes bibliographical references.
ISBN 0-660-13445-4 .
DSS cat. no. R61-2/9-46E

1. Roofing, Slate. 2. Shingles. 3. Roofing — Canada — History. 4. Building materials — Canada. I. Canadian Parks Service. National Historic Parks and Sites. II. Title. III. Series.

TH2445.C87 1990   695   C90-098518-6

Contents

Acknowledgements 5

Introduction 7

  I. A Late 19th-century Urban Phenomenon 9

 II. Stylistic Trends in the Use of Slate Roofing in Canada 27

III. Canadian Practice in Laying Slate Roofs 45

IV. Sources of Roofing Slate, Past and Present 55

 V. Conclusion 67

Illustrations 71

Endnotes 137

Selected Bibliography 155

Submitted for publication in 1984 by Mary Cullen, Architectural History Division, National Historic Parks and Sites, Parks Service, Environment Canada, Ottawa.

Acknowledgements

Research for this study tapped the resources of many archives and libraries. I thank the staff of the National Archives of Canada, the Canadian Inventory of Historic Building, libraries of the Canada Centre for Mineral and Energy Technology and the Geological Survey of Canada, the Archives of Ontario, the London Public Library, the Historical Society of Pennsylvania and the Wilbur Collection of the University of Vermont.

Many individuals and societies also generously shared with me their knowledge and unpublished research. I am indebted particularly to Keith Blades, Gail Sussman, Martin Weaver, Gordon Phillips, and Elizabeth Vincent, all of Ottawa; Barbara Salomon de Friedberg, Québec; the Sisters of the Congregation of Notre Dame, Montreal; La Société d'histoire des Cantons de l'Est, Sherbrooke; Elizabeth Spicer, London, Ont.; Ann Gillespie, Dundas, Ont.; Trinity Historical Society, Trinity, Newfoundland, and Philip Marshall, Burlington, Vermont.

The slate industry provided invaluable assistance. I am very grateful to Bedford Slate Ltd., Montreal, the Stoddard Slate Companies, Bangor, Pennsylvania, the Buckingham-Virginia Slate Corporation and especially to William M. Mahar of the Vermont Structural Slate Company, Fairhaven, Vermont, who provided samples of current American roofing slates.

Last but not least, I thank my colleagues at the Architectural History Branch and Restoration Services, Canadian Parks Service.

All illustrations and photos are by the Canadian Inventory of Historic Building (CIHB), Environment Canada, or by the Heritage Recording Services (formerly of Environment Canada and now with Public Works) unless otherwise stated. Figures with asterisks are colour photos.

Introduction

As a roofing material slate has historically been unparalleled in its combined attributes of permanence and appearance. A natural building stone, this metamorphic rock, characterized by an excellent parallel cleavage, possesses the essential mineral constituents white mica and quartz that are highly resistant to water absorption and weathering or abrasion. The varying presence of accessory constituents, such as hematite and chlorite, has produced slate in a variety of desirable colours. Yet, as this study indicates, being a natural building stone also limited the use of slate as a roofing material. Even with the most efficient quarrying methods serviceable slate has always represented as little as 10 to 40 percent of material quarried. Slate is not only expensive to produce, it is heavy. The weight of slate increases freight charges and renders installation difficult in comparison to factory-made materials. For these reasons, in this century slate ultimately failed to compete with manufactured roofing products.

Slate roofing flourished in Canada in the last half of the 19th century stimulated by architectural fashion, the railway and urban development that made quarrying cost-effective. Slate roofs continued to appear between 1900 and 1930. But in these decades the market for slate, increasingly assaulted by cheaper products, was reduced to the wealthy client. This book first discusses the rise and decline of the material indicating where it was most popular and why. With a view to suggesting the most suitable treatment in historic buildings, the range of decorative possibilities in slate are then described and related, by example, to changing architectural styles. Finally, on a practical level, I review how Canadians usually applied or installed roofing slate and where they obtained supplies of the material.

The specific purpose of this building material book is to assist preservationists, inside and outside the Canadian Parks Service, in the evaluation and treatment of building components from an historic "Canadian" perspective. This slate study and others in process, such as milled and finished lumber, will establish context of material use — time, place, stylistic interpretation and installation techniques. It is also hoped that they will suggest new approaches for looking at Canadian architecture and increase our understanding of how material choice and handling distinguish Canada's built heritage. Indeed, it may well be through the study of building materials and systems that historic Canadian ingenuity and innovation in architecture will be revealed and defined.

The research for this book drew on a wide range of documentation. The analysis of historic consumption and demographic trends in the use of roofing slate was largely based on annual reports of Canadian government departments and their branches — Geological Survey, Mines, Trade and Commerce — and supplemented by secondary sources and the architectural files of the Canadian Inventory of Historic Building (CIHB). Decorative effects in slating were defined from 19th-century encyclopedias and stylistic trends identified by sampling architectural treatises, popular pattern books and photographic archives. Information on slating methods was gathered from specifications, handbooks and professional periodicals. Details on the origin and types of roofing slate used in Canada were found in the Geological Survey Reports of Canada and the United States.

I. A late 19th-Century Urban Phenomenon

Slate roofing reached its greatest popularity in Canada during the 1880s. Usable domestic slate for roofing was discovered in the 1840s but the practice of slating here and elsewhere was never just a function of supply. In Wales, where the material was always plentiful, slate did not achieve its status as a national roof covering until mid-19th century when transportation improved, population and cities grew, demand for building material increased and architectural styles suited to slate roofing were popularized through pattern books and builders' manuals.[1] In Canada, railway development, city growth and architectural demand converged slightly later to produce the widespread use of slate. Slate consumption statistics pinpoint 1889 as the high point for the use of slate roofing in Canada. Extant slated buildings identified through the Canadian Inventory of Historic Building (CIHB) reinforce and amplify the statistical and historical record of slate roofing as a late 19th century, urban and, given settlement patterns, eastern phenomenon.

Slate is a fine-grained rock formed from clays and shales by intense metamorphism. The pressure and high temperatures of the metamorphic process rearrange the mineral grains of shale into parallel positions with a tendency to split with ease in one direction. Slaty cleavage, as this tendency is termed, produces thin, smooth, even sheets of rock which can be used for roofing. A proper roofing slate contains very little of the original clay but consists mainly of mica, quartz and chlorite, minerals that are highly resistant to absorption and weathering. Variations in mineral and chemical composition determine slate colour which ranges from blue, black, and grey to green, purple and red.[2] Slate, in sum, is a natural quarried product whose historic use depended on the development of markets able to support its cost of production.

The religious architecture of New France provides the first examples of the use of slate roofing in Canada. The initial churches, stripped down imitations of the monumental classic architecture of 17th-century France, used foreign building materials. The 1666 Jesuit church built on the Place du Marché at Québec, and designed in the form of a latin cross with a bell tower at the transept crossing, had a French-slated hipped roof.[3] A second example was Notre-Dame-de-l'Immaculée-Conception established as the parish church of Québec in 1664.[4] Seminaries and convents also used French slate. At Québec, the Jesuit College was slate covered and the second Ursuline Convent was partly roofed with slates. Accounts indicate that the nuns bought 38 000 slates in 1674 and paid *Maître* Robert Pepin "piqueur d'ardoise" for installing both slates and shingles.[5] In 1687, 36 000 slates, with 300 feet of lead and 60 000 nails were sent to Montreal from the port of La Rochelle in France to cover the Sulpician seminary then being built.[6]

French slate also covered several of the public buildings of the civil administration of New France. Inadequate slate supplies meant that several materials were frequently used on the one roof. In 1716 the Intendant's Palace was roofed with both boards and slate.[7] The King's Engineer Chaussegros de Léry wrote the Conseil de Marine in 1721 urging double the amount of slate be sent to the colony in order to cover the boarded King's buildings at Québec. "Roofs in slate last a long time," he noted, "and the repairs are not as big as for a shingle roof that requires frequent replacements and repairs"[translation].[8] The use of slate roofing for official buildings in Québec was observed by traveller Peter Kalm who recorded in 1749, "The roofs of the Public buildings are covered with common slate which is brought from France because there is none in Canada."[9] French slate was also installed on the barracks of the King's Bastion built at Louisbourg in the 1720s.[10]

Both legal and practical steps were taken by the authorities of New France to extend the use of slate roofing to private buildings particularly in the towns. After a fire which destroyed 138 buildings in Montreal in 1721, Intendant Michel Bégin issued an ordinance that all houses there should henceforth be built of stone and covered with a double covering of boards "until it is possible to use tiles or slate." A comprehensive building code for all the towns in the St. Lawrence Valley, issued in 1727, forbid shingles, specified roof framing and again recommended slate or tile covering.[11] Between 1728 and 1733, efforts were made to open a roofing-slate quarry at Grand Étang on the south bank of the St. Lawrence River between the seigneuries of Rimouski and Trois Pistoles. Samples sent to France were judged of good quality. De Léry himself covered his house with the local slate and Intendant Hocquart bought 101 600 slates on the King's account for roofing the palace and powder magazines. These trials soon revealed that the Grand Étang slate had poor cleavage and its rough surface was too porous to resist the elements. Since the poor local slate cost more than the superior imported product, the quarry was abandoned.[12] Despite the efforts of the 1720s, the practice of slate roofing gradually stopped even for church and public buildings whose owners could afford to buy foreign slate.

Whether because of the failure of the local industry or the expense and inadequate supply of imported slate, during the 1730s and 1740s, tinplate or *fer blanc* was imported for trial on the Intendant's Palace.[13] Documents record that tinplate, another incombustible material, had already been used in the colony as early as 1670 by the Ursulines and Sulpicians.[14] When the palace experience indicated that the tinplate soldering cracked when frozen, Chaussegros de Léry, the champion of slate, seized the occasion for its promotion: "Of all the roofs, the best in the Country is that in slate which resists the frost"[translation].[15] The re-covering of the palace with tin, despite this endorsement, hinted official disapproval or impatience with slate. One can only speculate on the use of slate for roofing if the French Regime had continued. After the British conquest the French architects and engineers who were slate's main promoters were either dead or had left the colony. Nearly a century would pass before slate roofing would again receive official recognition and acceptance for ecclesiastical and public building.

By the time English architects began to establish themselves in Quebec in the 1790s, tinplate had become the most popular fire-proof roofing material for churches. Both Lady Simcoe and Isaac Weld travelling in Quebec in the 1790s observed that tin roofing on churches was already "the custom of the country."[16] The "fer-blantiers" of New France had developed a method to prevent rusting by laying the squares diagonally and folding corners over the nails,[17] a practice continued by British builders in the construction of the Anglican cathedral at Québec between 1799 and 1804. English use of tin for roofing in Québec may have been reinforced by the fact that the chief supplier of tinplate in North America during this period was Great Britain.[18]

In the first half of the 19th century tin roofing also spread to commercial and public building to the exclusion of slate. An agent of the Phoenix Fire Insurance Company wrote from Montreal in 1809 "in the covering of houses in this country they never employ Tiles or Slates, which are not thought to stand the Climate, as well as Tin or Sheet Iron."[19] In its recommendations on the form and nature of building best adapted for barracks and storehouses, the 1825 Smith Commission on the defence of British North America reasserted the prevailing belief in the superiority of tin or iron over slate roofing:

The roofs of the principal buildings in this country are covered with plank, upon which sheets of Tin or Iron are nailed. We think this an excellent sort of roof, much less liable to be out of order and to require repair, than slate or tiles. We are of opinion that the roofs of all Barracks, Storehouses or Government Buildings in the Canadas ought to be covered in this way.[20]

Contrary to British military practice in Quebec, the Atlantic colonies of British North America during this period saw general use of slate for stone public and commercial buildings. In her study of roofing materials used by the Royal Engineers in Canada, Elizabeth Vincent has found that while spurning slate in the upper colonies the Engineers preferred the material on the east coast. Duchess slates (slates 12 inches by 24 inches) were adopted for covering the magazines of the Halifax citadel and its dwelling casements and the original slates of the south magazine covered in the 1840s are still intact.[21] Many Maritime public buildings were roofed with slate: Province House, Halifax, 1811-18; Admiralty House, Halifax, 1819; Province House, Charlottetown, 1843-47; and Government House, St. John's, 1827.[22] The major seaports of Saint John, Halifax, and St. John's were predominantly wood built, shingle-roofed towns. However, through the century numerous fires eventually made brick or stone buildings roofed with slate the norm for waterside warehouses and stores.[23]

At mid-century slate was still a forgotten roof covering in central Canada. In the mercantile part of the city of Montreal which included "the most elegant and substantial of the public Buildings" nine-tenths of the buildings had roofs of tin or sheet iron. Insurance reports pointed out that "Tiles or Slates being nowhere used in the Colony."[24] The prevailing disregard of slate was reiterated in a *Quebec Gazette* article on various kinds of roofing used in Lower Canada. The feature on 10 June 1846 reported, "Slates are at present but little appreciated in this country although they have stood the test of climate."[25] The negative attitude toward slate roofing in pre-Confederation Ontario and Quebec is most practically and quantitatively illustrated by G.F. Baillairgé's 1867 report describing the public buildings of Upper and Lower Canada. The document which summarizes most major construction since 1830 lists about 115 buildings including houses of parliament, observatories, custom houses, post offices, court houses, jails and drill sheds. Four buildings have slate roofs, the most significant being the houses of parliament just completed in the capital of the new Dominion of Canada.[26]

The choice of slate for roofing the Ottawa Parliament Buildings, a construction project of unprecedented magnitude in British North America, was the result of several circumstances that dramatically changed builders' attitudes toward the use of this material in the 1850s and 1860s. The population of Upper and Lower Canada almost tripled from 637 000 in 1825 to 1 842 000 in 1851.[27]

The rapid settlement stimulated financial and commercial development of Toronto and Montreal, and created an increased demand for building materials of all kinds. Railway improvements contributed to the urban growth and made possible commercial slate quarrying in the United States and Canada by linking cities with quarries away from ship routes.[28] Henceforth a supply of roofing slate was readily available and reasonably priced. Finally a shift from conservative Neoclassical public building that de-accentuated rooflines to the Gothic Revival and Second Empire styles featuring steeply pitched, coloured, patterned roofs encouraged the use of slate as it could be readily fitted to complex roof forms and lent itself to fanciful designs.[29]

Beginning in the 1850s Canadians had access to closer and cheaper sources of roofing slate than those historically provided by European and British quarries. Slate quarrying had taken place in the northeastern United States since 1734 but only started to be exploited on a large scale in the 1850s.[30] Railway and urban development in Canada encouraged the export of United States roofing slate north, a pattern of trade that was further supported by the Reciprocity Treaty of 1854 which exempted American slate from duty. Tables of the Trade and Navigation of the Province of Canada for the period of the treaty from 1854 to 1866 (Table 1) show that, except for three years when a small portion was admitted from Great Britain, all the slate imported into Canada came from the United States.[31] The value of the American slate imported under the Reciprocity Treaty reached a high point of $12 763 in 1859, its subsequent decline reflecting the disruptive effect of the Civil War on American production and the development of domestic sources in the 1860s.

The Geological Survey of Canada identified localities in the Eastern Townships of Quebec (Fig. 1) suitable for quarrying roofing slate as early as 1847.[32] Specimens of roofing slate from Kingsey, Frampton, Melbourne, Shipton and Tring were shown at world industrial exhibitions in London (1851) and Paris (1855).[33] Quarries operated briefly at Shipton and Kingsey in the 1850s but the first viable enterprise was the Walton Slate Quarry (Fig. 2) established in the township of Melbourne in 1861.

A variety of sizes was produced to sell at an average price of $3.80 a square (the amount needed to cover 100 square feet of roof surface with a three-inch lap). This price provided roofing slate in Canada at or under London prices[34] and prompted the use of Melbourne slates in the covering of the Parliament Buildings then under construction.[35] In 1868, a second slate quarry was established two miles away by Montreal entrepreneur Charles Drummond. The New Rockland Slate Quarry soon surpassed the Walton company producing in 1875, 7000 to 8000 squares of roofing slate in contrast to 3000 squares by its competitor.[36] Roofing slate quarries were opened intermittently in British

Columbia, Nova Scotia and Newfoundland. The Newfoundland roofing slate industry, successful for a brief period from 1902 to 1905, sent most of its product to England.[37] The two Eastern Townships quarries, which began production in the 1860s, were always the principal source of roofing slate production in Canada and supplied most of the needs of the Canadian market until 1900.

The simultaneous beginning of a domestic roofing slate industry and the adoption of Gothic Revival and Second Empire styles for public building in the new Dominion were pivotal in increasing the popularity of slate roofing in Canada. The 1867 Parliament Buildings set the tone, embodying design elements of both styles and emphasizing the picturesque effect of the roof by the use of polychromatic (variously coloured), slate-covered mansard roofs and towers (Fig. 3). After Confederation the federal Department of Public Works chose the Second Empire Style to spread the federal image in an ambitious building program for post offices and custom houses throughout the country. The identifying feature of this style was the mansard or broken roof with slate arranged in decorative patterns on the slopes.[38] In contrast to the four slate-roofed buildings listed in the 1867 Public Works report, Chief Architect Thomas Fuller's report on public buildings for 1867-82 records 26 with slate roofs.[39] "Slates, 20 in. by 10 in., of approved tint or quality, from Richmond," was a standard phrase in specifications for federal post offices and custom houses (Table 2) until the late 1880s.

Enthusiasm for Second Empire, fanned by federal allegiance to the style, was also manifest in public architecture at the provincial and municipal levels, in commercial and religious buildings and in the homes of wealthy fashion-conscious urban residents. Important design features such as pavilions and rich sculptural decoration were often lost in the diffusion of the style but the visually alive mansard roof remained a characteristic trait. During the 1870s "mansardization" in fact became a craze, frequently being superimposed on buildings whose main stylistic roots lay elsewhere.[40] In central Canada and particularly in Ontario (see next chapter), mansard roofs were covered with variously cut and coloured slates.

Slate roofs were also a basic element in the Canadian treatment of the High Victorian Gothic style which emerged in the second half of the 19th-century. Abandoning the rational structuralism which dominated earlier phases of the Gothic style, architects embraced greater freedom of form with exuberant visual effects. In the execution of the irregular silhouettes, vertical thrust and colour accents that distinguished this free interpretation of the Gothic Revival,[41] roof form and covering played an essential role. University College, Toronto (1856-59) "one of the first great High Victorian Gothic Buildings in Canada,"[42] was capped by a banded polychromatic slate roof (Fig. 4). A similar slate pattern was repeated with even greater visual effect on the Ottawa River elevation of the Parliament Buildings (Fig. 5). The fashion was quickly assimilated by religious and domestic architecture creating a wider market for slate. The consumption of slate in Canada peaked in 1889 (Tables 3 and 4) just before the influence of the High Victorian Gothic began to fade in the 1890s.

The principal periods for the popularity of slate roofing in Canada can be identified precisely by a review of the annual values of locally produced and imported roofing slate that appear in the reports of the Division of Mineral Statistics and Mining of the Geological Survey of Canada and the Tables of the Trade and Navigation of the Dominion of Canada. Canadian slate statistics, which begin only in 1886, are slightly skewed by the inclusion of school blackboard slates in production figures from 1886 to 1896. By ascertaining the value of school slates from other sources, however, it has been possible to isolate the value of roofing slate production. Table 3[43] showing annual consumption on a monetary scale for the 50-year period from 1886 to 1936, indicates slate roofing enjoyed greatest popularity from 1886 to 1894, 1906 to 1914, 1921 to 1923 and 1928 to 1930. Quantities of imported and Canadian roofing slate (Table 4)[44] that would provide a more accurate index of use are not recorded in comparable units before 1900. Though Canadian production is initially recorded in tons, an 1889 Geological Survey report stated that the New Rockland Slate Quarry, the only quarry in operation that year, produced 26 400 squares of roofing slate.[45] Added to the import of 8417 squares of American slate, Canadian consumption in 1889 thus attained 34 817 squares. Between 1906 and 1914 when values again rose Canadians annually used between 20 000 and 22 000 squares of slate. About a third of this quantity was used from 1928 to 1930, although the total value or price of slate exceeded pre-war figures. Both yearly values and quantities of roofing slate thus reveal that the popularity of slate roofing in Canada reached its zenith in the period from 1886 to 1894. So great was the demand for slate in 1889, roofers complained to the *Canadian Architect and Builder* that the duty must come off foreign slate since the New Rockland Company could not provide the necessary supplies.[46]

Five years after the market for roofing slate appeared insatiable, thirst for the material abated not to revive until after 1900. Technical advances, economic conditions and architectural taste all seemed to be at play in the decline, illustrating the difficulty of isolating factors influencing the use of construction materials. A great reduction in the price of copper about this time[47] may have induced the federal government to reconsider the advantages metals offered in light weight, incombustibility and easy installation. After 1888, Public Works specifications began to call for galvanized iron and copper roofing rather than slate.[48] Copper roofing replaced slate on the front of the

Centre Block in 1890 and also covered the Langevin Block constructed between 1888 and 1890.[49] Reaction to the picturesque forms of High Victorian Gothic and Queen Anne Revival architecture also set in during the 1890s. The eclecticism of the era continued in the parade of Beaux-Arts, Château and Romanesque designs that began to appear in Canadian cities, but the former polychrome effects in roofing were now passé. Many architects and builders working in these styles found metal and other roofing materials acceptable where hitherto only slate had been able to provide desired patterns and natural variety of colour. Besides the rivalry of other materials and changing taste, world recession contributed to the decline in the use of roofing slate in the 1890s. From a high of 200 men in 1893, the workforce of the New Rockland Quarry was reduced to 90 in 1897.[50] Canadian production temporarily stopped in May 1900. Although it resumed a year later, the quarry never again supplied more than a third of the Canadian consumption of roofing slate.[51] (See Table 4).

Slate roofing enjoyed a second wave of popularity in Canada from 1906 to 1914, the result of one of the most vigorous periods of industrial growth in our history. As manufacturing concentrated in cities, the urban population of Canada increased from 29.8 percent in 1881 to 41.8 percent in 1911.[52] The settling of the west and boom in railway construction further stimulated eastern secondary industry and urban growth. New municipalities sprang up on city peripheries, many of them later being annexed to the larger centres. Building in the multiplying middle- and working-class neighborhoods exhibited a type of economic construction and a high degree of industrialization. Flat roofs which cut expenses in time and money became common but many architects still resorted to an 80 degree slate slope of four to six courses above the cornice to improve the design. In some planned industrial towns such as the present Montreal area of Maisonneuve, row houses were required to have stone façades and slate roofing continued to be employed.[53] Thus slate in this period was not just evident on institutional architecture and wealthier houses, but also in the common domestic buildings the row houses and duplexes that lined the expanding working-class sections of Canada's urban centres.

The pre-war boom in building construction which stimulated the use of slate roofing was ultimately its undoing. Architects, now concerned about the speed of construction and the cost per square foot of space, began to demand roofing materials that were cheap, easily handled and installed. In 1887 the *Scientific American* hailed the introduction of sand-coated fibrous pulp or felt roofing as "superior to that of slate because of its lightness."[54] Since then a whole range of materials from cement, asbestos and even crushed slate was developed as surfacing for asphalted felt and marketed as prepared or "ready" roofing. Slate trade literature reacted to the new competition. In its 1907 brochure entitled "Slate and Its Uses" the Genuine Bangor Slate Company of Easton, Pennsylvania, pressed the advantages of slate roofing in comparison to standard competitors such as tin, steel and corrugated iron. But it launched its lengthiest and most vigorous defence against composition roofing, manufactured coverings made of tar, asphalt, gravel or asbestos coatings on felt, wool and other materials.[55] The *Contract Record*, sequel to the *Canadian Architect and Builder*, declared in August 1911 that "prepared or ready roofings have come to stay."[56] By the 1920s asbestos and other composition shingles were being marketed in a variety of shapes, sizes and a wide range of colours.[57]

To understand why slate failed to compete with the newer manufactured roofing materials, many of which were inferior to average slate in attractiveness and permanence, one must look at the nature of the respective products. Factory-made materials were subject to the economics of mass production and to a high degree of mechanization that lowered costs. As a natural product, roofing slate was deposited in specific areas or beds of a rock structure, the serviceable stone representing as little as 10 to 40 percent of the amount quarried. Mechanization of slate quarrying by the introduction of the track channeler (1897) and the wire saw (1926)[58] reduced some industry waste, but was unable to overcome the cost edge of the synthetic materials. Added to the inherent expense of producing roofing slate was the weight factor. Since slate was a heavier material than most of its competitors it had to bear the cost burden of higher freight charges and the inconvenience of handling. Finally the proper setting of slate on a roof demanded more skill than that required by certain other roofing materials, a degree of care that was not appreciated by speculative builders.[59]

In the 1920s and 30s, slate roofing appealed to a restricted section of the building industry conscious of capitalizing on the design potential of slate and its reputation for permanence. During this period architects developed great enthusiasm for coloured, thick heavy slates known as architectural grades to achieve various effects ranging from the simple wood shingle look to the antique graduated roof. Slates two-inches thick and weighing from 75 to 200 pounds each were popular for roofing large residences. Special demand was created for the red slates of New York and the variegated green and purple slates of Vermont. The price per square of the architectural grades, as much as $50 per square for 1 1/2-inch-thick red New York, suggested that slate had become the roofing material of the wealthier client who could afford an architect designed home.[60] In 1930 the value of roofing slate used in Canada reached $118 765, nearly $17 000 more than its pre-war value, yet the quantity was little more than a third that consumed in 1914. (See Table 4 here and Table 5 in Chapter 4).

Since 1932 the use of slate as a roofing material has been negligible in Canada. Less than 2000 squares of roofing slate have been imported annually. (See Tables 4 and 5). After roofing slate production ceased in Quebec in 1921, slate quarries there and in British Columbia produced granulated slate for surfacing asphalted roofing paper.[61] Urban growth, which had originally abetted the use of roofing slate, ultimately produced a market for cheap building materials that could not be satisfied by costs of producing slate, whether imported or domestic.

The approximately 3000 slate-roofed buildings identified by the Canadian Inventory of Historic Building computer program confirm the historic profile of slate roofing as a 19th-century urban phenomenon. In the absence of historic marketing data, the inventory data also provide the best impression of the geographic distribution of slate roofing that we have found. Fewer than 60 slated buildings were recorded west of the Ontario border, a factor consistent with a region whose urban growth occurred after the heyday of slate. Nearly four times as many slate-roofed buildings were recorded in Ontario than in Quebec, while Newfoundland showed more than the other three Atlantic provinces combined. In Quebec, buildings with slate roofs appeared principally in Montreal and towns of the Eastern Townships such as Stanbridge, Richmond, Sutton and Danville. Samples in Québec were conspicuously absent. Half the slate roofs in Ontario were recorded in Toronto, the roughly 1000 structures there providing the largest concentrated sampling in the country. Sizable groupings also appeared in London and Brantford. In Newfoundland, slating was confined to St. John's and the Smith Sound area of the Avalon Peninsula.

The geographic distribution of slate roofing in eastern Canada can only partially be explained by urbanization and accessibility to materials. Slate roofs in the regions of Trinity Bay, Newfoundland and Eastern Townships, Quebec, certainly fit the classic reason for the use of materials — nearness to sources. Yet while most centres in southern Quebec and Ontario were about equidistant from Canadian and American slate quarries, there existed a wide discrepancy in the tendency of each to use slate. The association of slate with particular styles that were unevenly spread across Canada sheds some light on the varied concentration of slate roofing. The large sampling in London, Ontario, for instance, may be attributable to that city's formative architectural phase occurring at a time when such styles as High Victorian Gothic, Second Empire and Queen Anne Revival encouraged slating.[62] The relationship of style and material was nevertheless far from predictable. Slate roofs were integral in the federal expression of the Second Empire style, but the mansard-roofed cottages dotting the shores of the St. Lawrence River were covered in tin.[63] As styles diffused, the building practices and labour forces of local areas modified the materials of the prototypes. Studies of the artisan classes of Quebec may yet prove that the limited use of slate in that province was due to the existence of a skilled force of "fer-blantiers" which reinforced the persistence of a building tradition in imported tin long after local slate proved more accessible and less costly.[64]

What class of Canadian buildings had slate roofs? Contrary to the impression suggested by documentary research that governmental, religious and mercantile structures constituted the largest groups with slate roofs, more than two-thirds of the slate-roofs recorded by the Canadian Inventory of Historic Building were residences. Concentrated in Ontario and Quebec, the 2000 dwellings that emerged from the survey emphasize in a dramatic way the popularity of slate in its heyday. Slate-roofed detached residences appear to have been most prevalent, although there were also many slate-roofed townhouses in Montreal and slate-roofed, semi-detached homes in both Montreal and Toronto.

The CIHB record also confirmed my hypothesis that slate roofing was used most on stone and brick buildings. Slate-roofed brick buildings were the common material mix in Canada as befitted the higher occurrence of slate structures in the brick country of Ontario. Yet in certain regions where slate was quarried nearby, slate replaced wood and asbestos shingling as the universal covering for all frame buildings ranging from homes to barns and sheds. This phenomenon was particularly evident in Newfoundland (Fig. 6). Another example was Danville, Quebec, a centre for the quarrying and manufacture of school slate in the 1880s and 1890s. Here, nearly all the buildings recorded by the CIHB, most of them frame structures, were roofed with slate.

The demographic portrait of slate roofing in Canada then is typically residential, eastern and urban, with greater concentrations in Toronto particularly and Ontario generally. Slate roofing is most commonly seen on brick and stone buildings, but in areas of local slate supply, slate covered every manner of building. Historically, slate roofing reached its zenith in Canada during the 1880s although its use became common in mid-19th century and continued until the 1930s. How slate was used for roofing showed little geographic variation but appears to have changed with successive currents of architectural development. The material itself defined the decorative option that became the basis of stylistic trends in slate roofing in Canada and elsewhere.

Table 1
Values and Sources of Roofing Slates Imported into Canada for the period of the Reciprocity Treaty 1854-66

Year	Source	Value
1855	U.S.	£ 7398.8.7
1856	U.S.	5000.11.10
1857	U.S.	4280.7.6
1858	U.S.	15.
	G.B.	3.51.1
1859	U.S.	(Cdn) $12763.00
1860	U.S.	3700.
	G.B.	665.
1861	U.S.	5058.
1862	U.S.	1819.
	G.B.	170.
1863	U.S.	1914.
1864	U.S.	1045. (half yr.)
1865	U.S.	5214.

Tables of the Trade and Navigation of the Province of Canada for the Period of the Reciprocity Treaty 1855. (Toronto: Stewart Derbishire and George Desbarats, 1856)

Table 2
Slate-roofed Federal Buildings, 1867-90,
based on a survey of extant specifications

Year	Building name and location	Type of slate
1871	Post Office, Ottawa.	Melbourne, Que.
1872	Post Office, Montreal	Canadian
1872	New Savings Bank, Saint John, N.B.	Bangor Countess
1874	Post Office, Saint John, N.B.	Welsh Bangor
1874	Alterations, additions, old Post Office, Toronto	(heavy Canadian
1875	Extension of Western block, Departmental Bldgs. Ottawa	from approved quarries)
1875	Examining Warehouse, Montreal	slate
1877	Post Office-Customs House, St. Jean, Que.	Richmond, Que.
1878	Post Office, Fredericton, N.B.	Richmond or Vermont quarries
1878	Post Office-Custom House, Brantford, Ont.	Richmond or Vermont quarries
1879	New Custom House, Saint John, N.B.	best black Bangor, Wales or Pennsylvania?
1881	Government Workshop & Sheds, alterations, Parliament Hill, Ottawa	same as original
1882	Marine Hospital, Saint John, N.B.	Canadian slate
1882	Post Office-Customs House, St. Thomas, Ont.	Richmond or equal
1882	Post-Office-Customs House, Woodstock, N.B.	Richmond or equal
1882	Post Office-Customs House, Brockville, Ont.	Richmond
1882	Post Office-Customs House, Cornwall, Ont.	Richmond
1883	Post Office-Customs House, Amherstburg, Ont.	best slate
1883	Post Office-Customs House, Barrie, Ont.	Richmond
1883	Post Office, Moncton, N.B.	best slate
1883	Post Office-Customs-Inland Revenue, Port Hope, Ont.	Richmond
1883	Post Office, Carleton, N.B.	Canadian

Table 2
Slate-roofed Federal Buildings, 1867-1890,
based on a survey of extant specifications, cont'd.

Year	Building name and location	Type of slate
1883	Post Office, Summerside, P.E.I.	Canadian
1884	Post Office-Customs-Inland Revenue, Galt, Ont.	Richmond or other
1884	Post Office-Customs-Inland Revenue, Berlin, Ont.	Richmond or other
1884	Post Office, Orangeville, Ont.	Canadian
1884	Post Office-Custom House, Bathurst, N.B.	Richmond
1884	Post Office-Inland Revenue, Newcastle, N.B.	Richmond
1884	Post Office-Customs House, Sorel, Que.	Richmond
1884	Post Office-Customs House, Amherst, N.S.	Richmond
1884	Addition to Customs House, London, Ont.	to match present
1884	Post Office-Customs House, New Glasgow, N.S.	Richmond
1884	Post Office-Customs House, Yarmouth, N.S.	Richmond
1885	Post Office-Customs House, St. Stephen, N.B.	Richmond
1885	Post Office, Peterborough, Ont.	Richmond
1885	Customs and Inland Revenue, Peterborough, Ont.	Richmond
1885	Post Office-Customs House, Baddeck, N.S.	Richmond
1885	Dominion Building, Charlottetown, P.E.I.	Richmond
1885	Post Office-Customs House, North Sydney, N.S.	Richmond
1886	Post Office, Hull, Que.	Richmond
1886	Infantry School, London, Ont.	Richmond
1887	Post Office, Gananoque, Ont.	Richmond
1888	Post Office, St. Jerome, Que.	Richmond
1888	Post Office, Joliette, Que.	Richmond
1888	Addition to Post Office, Cobourg, Ont.	slate to match
1889	Post Office-Customs-Inland Revenue, Fraserville, Que.	Richmond
1890	Addition to the Supreme Court Buildings, Ottawa	slate equal to present

Canada, Department of Public Works, Specifications for Public Buildings, 1871-1907

Table 3
Values of Domestic and Imported Roofing Slate and
Total Consumption, 1886-1936

Year	Canadian slate $	Imported slate $	Totals $
1886	64 675	1798	66 473
1887	89 000	3002	92 002
1888	70 689	13 884	84 573
1889	99 160	33 254	132 414
1890	78 000	15 863	93 863
1891	65 000	34 584	99 584
1892	69 070	34 443	103 513
1893	90 825	26 952	117 777
1894	51 550	12 568	64 118
1895	43 400	5726	49 126
1896	38 850	9948	48 798
1897	42 800	5032	47 832
1898	40 791	3577	44 368
1899	33 406	9096	42 502
1900	12 100	22 278	34 378
1901	9980	38 009	47 989
1902	19 200	37 390	56 590
1903	22 040	44 573	66 613
1904	23 247	38 245	61 492
1905	21 568	45 345	66 913
1906	24 446	60 054	84 500
1907	20 056	51 826	71 882
1908	13 496	72 588	86 084
1909	19 000	62 132	81 132
1910	18 492	72 842	91 334
1911	8248	68 728	76 976

Table 3
Values of Domestic and Imported Roofing Slate and
Total Consumption, 1886-1936, cont'd.

Year	Canadian slate $	Imported slate $	Totals $
1912	8939	85 031	93 970
1913	6444	95 222	101 666
1914	4837	96 705	101 542
1915	2039	81 212	83 251
1916	6223	32 703	38 926
1917	7789	20 455	28 244
1918	5124	19 903	25 027
1919	10 853	47 268	58 121
1920	14 200	32 555	46 755
1921	22 325	71 341	93 666
1922	14 871	78 816	93 687
1923	17 289	70 298	87 587
1924		71 201	71 201
1925		62 906	62 906
1926		52 514	52 514
1927		64 698	64 698
1928		79 091	79 091
1929		87 804	87 804
1930	3000	118 765	121 765
1931	5000	56 578	61 578
1932	3750	41 376	45 126
1933	3750	8406	12 156
1934	4802	8724	13 526
1935	4329	12 150	16 479
1936	5414	11 547	16 961

Records of the Minerals Resources Branch, Slate Production Information, 1885-1919.
(NA, RG87, Vol. 34, No. 140).

Table 4
Quantities of Roofing Slate Produced in Canada
and Imported, 1886-1936

Year	Canadian production	Imports
1886	5345 (tons	326 (squares)
1887	7357	543
1888	5314	3184
1889	6935	8417
1890	6368	3207
1891	5000	7939
1892	5180	15 552
1893	7112	12 816
1894	data not available	3067
1895	data not available	1470
1896	data not available	2891
1897	5208	1566
1898	3432	1540
1899	data not available	4407
1900	3170	8018
1901	715	11 678
1902	4800 (squares)	11 319
1904	5277	8714
1905	4900	10 920
1906	5469	15 373
1907	4335	14 402 (9 months)
1908	2950	17 734
1909	4000	15 079
1910	3959	18 767
1911	1833	16 919
1912	1894	20 213
1913	1432	21 457

Table 4
Quantities of Roofing Slate Produced in Canada and Imported, 1886-1936, cont'd.

Year	Canadian production	Imports
1914	1075	21 702
1915	397 (squares)	16 394 (squares)
1916	1262	7207
1917	1422	4328
1918	933	3716
1919	1632	8144
1920	no production	4694
1921	6086 (tons granulated slate	6662
1922	1899 for roofing paper)	6095
1923	1836	7028
1924		6076
1925		5033
1926	no production	4512
1927		5297
1928		7053
1929		6986
1930	150 (tons)	8941
1931	250	4347
1932	250	3687
1933	250	1192
1934	738	1135
1935	1129	1236
1936	1247	1425

Compiled from:
Canadian Mineral Statistics 1886-1956; Mining Events 1604-1956, Reference Paper No. 168 (Ottawa: Queen's Printer, 1957), and *Tables of the Trade and Navigation of the Dominion of Canada*. (Canada, Dominion Bureau of Statistics, Industry and Merchandising Division, General Statement ... Roofing Slate).

II. Stylistic Trends in the use of Slate Roofing in Canada

The basic decorative possibilities of slate had been established centuries before their various manifestations in the architectural fashions of late 19th- and early 20th- century Canada. Simply, the appearance of a slate roof might vary according to the nature of the material itself or the way it was cut and applied. Size, thickness, colour and cut were four main factors that determined effect. Around one or more of these features developed the trends in slating associated with different architectural styles in Canada from the mid-19th century to the 1930s.

Slate size refers to the rectangular size, split and trimmed from the quarry slab. Historically it ranged from as small as 11 inches by seven inches to as large as 36 inches by 24 inches. In Britain common slate sizes were given specific names, many of them royal:[1]

Name	Length	Width
Doubles	13"	6"
Ladies	16"	8"
Countesses	20"	10"
Duchesses	24"	12"
Welsh rags	36"	24"
Queens	36"	18"
Imperial or Patent slates	24"	30"

American and Canadian slates were categorized by dimension and covered the gamut of British sizes with the exception of the large imperial or patent slates.[2] French slates were usually not larger than the British doubles.[3] It was the part of the slate exposed on the roof, known as the margin or gauge, that determined the effect. In speaking of slate size, the method of application was as important as the quarry-produced size because the margin or exposed part could be anywhere from a third to nearly half the slate length depending on the lap used. Assuming an average lap of three inches, the standard French slate roof showed four inches to the weather while the British "Duchess" slate exposed a margin of 11 inches. The decorative impact of small slates was to increase the scale of the roof. Practically, the choice of slate size was limited by roof pitch and the scale of the building. The general rule according to 19th-century encyclopedias of architecture was the steeper the pitch and more exposed the position the smaller slating should be. Conversely a low-sloped roof required large slates, but not if the building was small.[4]

The thickness of slates was a second factor which could substantially alter the look of a roof. A quarried block of slate was split with hammer and chisel along the cleavage planes into individual laminae or slates. The thinner the laminae the better the quality of the slate. The best cleavage usually occurred in the darker coloured black, blue and grey slates which were split in sizes about 3/16 of an inch. Some Welsh slates were as thin as 1/32 of an inch, but these were not recommended for reasons of breakage. Coloured green, purple and red slates were ordinarily coarser, splitting to about 1/4-inch thick or more.[5] Architects who favoured texture on a roof preferred the heavier coloured slates instead of the thinner, smoother "blue" slates which they compared to metal. Before the mass production of slate and standardization of thicknesses, builders commonly sorted roofing slate placing the thickest at the lower edge of the roof, the medium towards the centre of the slope and the lightest at the ridge. To achieve the plastic or sculptural appearance of such roofs, 20th-century architects specified slates of diminishing thicknesses for what they labelled the "graduated" slate roof.[6]

A third decorative determinant in slate roofing was colour. Slate colours resulted from variations in mineral composition: chlorite in the greens, hematite in the purples, and iron oxide and hematite in the reds.[7] Specific quarries were usually identified with a particular colour from the grey slates of Angers, the purple of the Ardennes, the blue slates of Bangor in Carnarvonshire and the esteemed bluish-green of Kendall in Westmorland. The study of slate sources (see Chapter 4) also shows several colours often occurred within the one quarry, or as Vermont's variegated purple and green slates indicate, within the one slate. This great variety of colours and shades created unlimited possibilities for colour effects from the uniformly coloured slate roof in cold or warm values to the constrasting multi-coloured or polychrome surface in lively hues or subtle shadings. To compound the choice, slates could be permanently coloured or "fading" respectively producing the unfading tone or a weathered appearance which, several years after installation, was radically different from its original colour. Even slate lustre, while not strictly colour, was used to obtain variation. The 19th-century architect, Viollet-le-Duc thus records how French slaters applied the metallic coloured slate of the Anjou region in diverse ways to produce patterns from the sun's reflection.[8]

The fourth and perhaps most ornamental variable in slate roofing was cut. Cut refers not to the rectangular size but to the shape of the tail or exposed edge of the slate. Slate shapes assumed many forms, some specific to the slating craft itself and others borrowed from the centuries old decorative arts repertoire that drew its inspiration from the feathers, palm-leaves and fish scales of nature. There were about six best known shapes (Fig. 7). *Plain* or *common* slating referred to tail edges chipped to a straight horizontal line parallel to the upper edge or slate head. Sometimes the corner angles of plain slates were cut away forming broken lines. When the tail was kept straight but the whole slate reduced in size to a geometric square and applied so that only half the square appeared above the slate under it, the result was *lozenge* or *diamond* form.[9] The lozenge shape was also obtained by cutting a sharp line to the middle axis of the tail and laying the slate so that the visible part formed a net of diamonds. If the angle formed with the centre was less acute the cut formed a *hexagon*. Slates were also *rounded*, an ornamental option that originated in the necessity of covering conical roofs without the leaky and unsightly protrusion of square corners.[10] Variations on this semi-circular shape ranged from the slightly pointed *American Cottage* to the more acutely pointed *Gothic*.[11]

All the slate cuts, except the plain, belonged to the decorative arts family of scales which refers to those ornaments composed of symmetrical plates overlapping each other. Imitating the scales that cover fish, scale ornamentation was used by almost every people and at all times by a great number of industries — "the roofer in terracotta, wood or slate, the sculptor, the cabinet-maker and joiner, the zinc and copper worker, the jeweller ...."[12] Bands, borders, squares and other decorative elements were secondary designs developed by the slate roofer and other artisans to break large scaly surfaces as well as accentuate and alternate different cuts. The quincunx (arrangement of five objects in a square or rectangle) and the band or strip of sawtooth scales (Fig. 8) were just two of the ancient decorative devices that reappeared in the slating fashions of the 19th century.

The different shapes or cuts of slate were marketed by American slate companies under specific trade names. The slate shapes available in 1857 from the Eagle Slate Company in Vermont were named and illustrated in a contemporary advertisement (Fig. 9). Shape names such as American Cottage, diamond and plain remained current for several decades appearing in other publications such as Stafford's *Slater's Manual* and a book on slate roofing by Auld and Conger of Cleveland, Ohio. New designations appeared too, such as Washington for the hitherto unnamed slates with corner angles cut away. The slate cuts were apparently made by the slater himself with hand tools or a special machine (also used for punching or holing slate) sold by the slate companies.[13] No Canadian trade literature naming cuts has come to light.

Ornamental scales in the Gothic Revival

The first expressions of the Gothic Revival in Canada, manifested in Gothic details grafted onto churches and houses of traditional classical forms, gave no particular play to the roof. Nevertheless the pointed or Gothic style raised the roof from its Neoclassic decline and re-established it as a visible part of architectural design. Roof visibility fitted the growing 1840s interest in structural rationalism in which each element in the plan would be expressed in the exterior composition.[14] Even as Britain's Pugin and the Cambridge Camden Society were applying these ideas to church design, at a more popular level American landscape gardener and architectural critic, Andrew Downing, was formulating a principle that gave new emphasis to the roof. In his widely published designs for various cottages and villas entitled *Cottage Residences* (1842), Downing stated:

> *The principle of expression of purpose, demands that roofs of buildings should be shown, and rendered ornamental. In snowy countries especially a moderately steep roof is necessary to sustain the pressure and shed the snow perfectly, and it should always, therefore, be boldly exposed and rendered ornamental in domestic architecture.[15]*

The decorative treatment which Downing recommended to break up the plainness of the steep roof was cutting the tails of wood shingles or slates to create patterns when laid. Diamond, round and hexagon were the three shapes illustrated (Fig. 10) in his *Architecture of Country Houses*. Downing's designs for cottages in the rural Gothic style and villas in the "bracketted mode" respectively showed a diamond-pattern pavilion roof and a round-shingled gable roof.[16] A combination of three or four courses of ornamental shingles between several courses of plain was also seen as desirable. Downing acknowledged that nine-tenth's of country houses in America were wood shingled, but his decorative advice also applied to slates for he wrote "in country houses of the first class, where slates are used for roofing, they may be cut in the same patterns as shingles."[17]

At mid-century the great English critic John Ruskin saw scale decoration on the northern or steep roof as an appropriate symbol of the waterproofing armour of the fish-scale.

On steep domestic roofs, there is no ornament better than may be obtained by merely rounding or cutting to an angle, the lower extremities of the flat tiles or shingles ...: thus the whole surface is covered with the appearance of scales, a fish-like defence against water, at once perfectly simple, natural and effective at any distance.[18]

Ruskin placed great emphasis on the decorative value of dormers, finials and fringes. Beyond the use of scales, he saw little additional need for further enriching roof ornamentation.[19] It would be left to others to translate his enthusiasm for polychrome masonry effects into colour arrangements in slate roofing.

Ornamental scaling of the type recommended by Downing and Ruskin was not widely implemented in slate roofs in Canada for slate at this stage was just becoming available. Yet studies of the Gothic Revival,[20] show that Canadian builders were informed of developments in the Gothic Revival and in some instances used diamond and round scale roof decoration in other materials. Ornamental slate scaling was identified by the CIHB in Gothic-inspired houses at Madoc and Brantford, Ontario (Figs. 11 and 12*). [Figures with asterisks are colour photos.] The round angled slates at Yates Castle in Brantford were typical of the early Gothic Revival but their arrangement in bands of black alternating with green and red signalled the arrival of colour which soon became the characteristic trait of slating in the High Victorian mode.

Polychrome patterns in High Victorian Gothic and Second Empire modes

About mid-19th century, an increased sensibility to colour in Gothic architecture was encouraged in new interpretations of the Gothic Revival by British architects such as William Butterfield and George Edmund Street and the writing of the influential aesthete John Ruskin. Butterfield's All Saints' Church, Margaret Street, London (1850-53) built of red brick banded and patterned with black brick, introduced polychromy into English Gothic Revival architecture just at the time Ruskin was urging the importance of colour and the study of Italian Gothic models.[21] In 1855 Street's *Brick and Marble Architecture in the North of Italy* focused attention on the possibilities of geometric mosaics in these and other materials.[22] Polychrome soon became the principal symbol of the new High Victorian Gothic,[23] and, coupled with the high pointed roofs, large scale and bold silhouettes that were also features of this phase, it opened the way for colour effects in many materials, including slate.

In Canada the taste for polychrome effects in slate roofing initially appeared in High Victorian Gothic institutional and governmental architecture. University College, Toronto (1856-59) was one of the earliest examples. Inspired in composition by the newly completed Oxford Museum,[24] the College (Fig. 4) incorporated Norman as well as Gothic elements and was crowned by mansard pavilions and a medium-pitched roof with wide bands of alternating slate colours. On the roof of the west elevation (Fig. 13) slates were worked into rows of zigzags and diamonds enclosed by parallel horizontal lines. Similar geometric patterns had recently been employed by Butterfield in the brick walls of All Saints' Church (Fig. 14). The diamond or quincunx had been used before in 14th-century French slating but in the new Gothic spirit, the banding, bordering and zigzags of University College recall medieval models, the diapering of brickwork and the marquetry of marble and tile.

Within a few years polychrome pattern slate roofing was used on the Parliament Buildings in Ottawa. The "noble civic buildings of the Low Countries and Italy" provided suggestions for the design, but the High Gothic interpreted here liberally mixed its models of inspiration producing a vigorous composition of mixed Gothic, Second Empire and other sources enriched by the colourful treatment of building materials.[25] The slate roof of Fuller and Jones' central Parliament Building was striped in contrasting bands of plainly cut yellow and green slates. Repeated a decade later on the conical tower of the chapter-house Library of Parliament (Fig. 5), the roof stripes capped a building complex unparalleled in its picturesque effect. In their design of the Departmental Buildings that flanked the legislature, architects Stent and Laver used similar slate colours but created a more whimsical roof pattern (Fig. 15) alternating bands of plain and hexagonally cut slates and punctuating the latter by rosettes. These ministerial offices were more informal in character and their polychrome floral slate patterns paralleled ornamental slating in contemporary domestic architecture.

Polychrome slating associated with the High Victorian Gothic mode was undoubtedly a stimulus in the first large-scale commercial exploitation of slate in the United States in the late 1850s. American design books, at any rate, were not long in linking the availability of coloured slate with suggestions for roof patterns in varying tints. In 1857 architect Calvert Vaux, a former colleague of the now deceased Downing, wrote in *Villas and Cottages*:

Lately, new American quarries, supplying slate of different colours, have been opened in various parts of the country and worked with

*success. The slate that come from the Eagle quarries in Vermont is of
two tints: the one a rich purple-gray, the other a delicate green. This
slate, when arranged on a roof in stripes or patterns, so that the colours
are equally represented, has a very agreeable effect, and one that is far
superior to that produced by any shingle or metal roof.*[26]

Similar endorsations of two or more slate colours for the same roof were
made by Samuel Sloan, Hudson Holly and other purveyors of Gothic home
designs.[27] The elaborate "picturesque Gothic cottage" illustrated in these
popular publications was rarely duplicated in Canada. However, the L-shaped
house with the increased vertical proportions of the last half of the 19th century
often employed polychrome slate roofing effects to complement or replace
mural colour arrangements (Figs. 16 and 17).

Appropriately enough, because it was in ecclesiastical architecture that the
High Victorian Gothic had its most enduring influence in Canada, churches,
more than any other single type of building, illustrate the polychrome slate
fashion. In her study *Gothic Revival in Canadian Architecture*, Mathilde
Brosseau documented the preponderance of High Victorian religious architecture in Ontario, a conclusion corroborated by CIHB. Its printout on slate
roofing indicated more Ontario churches with polychromatic slate roofing that
churches in any other province of Canada. Ontario's architects obtained
commissions from various denominations so the practice of polychrome slating crossed religious and regional boundaries. St. Michael's Roman Catholic
Church (1845-66) and its neighbour the Metropolitan Methodist Church
(1872) in Toronto (Fig. 18) provide striking examples of the diversity of
patterns that were executed in slate. Delicate slate borders and decorative
crosses embellish St. Michael's roof while spikey elongated diamonds vigor-
ously zigzag their way across the main roof of the Methodist Church. The
ecclesiastical icons of the former example appear rarely in slate roofing
although the "Cross Patonce"[28] of the 1880 Central Presbyterian Church in
Cambridge (Fig. 19) provides another charming exception. The diamond-slate
pattern of the Toronto Methodist example was ubiquitous and various, forming
simple horizontal rows of large or small diamonds (Figs. 20* to 22), or
diamond rows enclosed with parallel lines or courses of round slates (Figs. 23
and 24).

The colour banding and zigzags illustrated in Cumberland's University
College were equally popular for churches. The striped composition of red and
grey rounded slates against a background of regularly cut black slates that
covered Henry Langley's Toronto Necropolis [1872] (Fig. 25*) was one of the
most striking designs of the period. Less colourful but more complex were the
alternating bands of grey and black slating which appeared on St. Peter's

Basilica, London [1880-85] (Fig. 26*) and the vigorous combination of horizontal banding and zigzag slating illustrated on St. Patrick's Church, Hamilton (Fig. 27). For all these patterns — the religious icons, the diamonds and zigzags, bordered or free-standing, and the contrasting colour bands — there were no specific formulas or recipes. There were no "pattern books" of slate roof designs for churches. Photos of extant slate roofs show that slaters used plain or round cuts, or layed slates diagonally to achieve the pattern, but colour arrangement was the key. Polychrome picked out the pattern, and was the trademark of slate roofing in the Gothic Revival churches of Ontario until the 1890s.

In Quebec and the Atlantic provinces polychrome slating was unusual in religious architecture. Part of the reason was the minor influence of the High Victorian Gothic in these regions. During the period the Quebec Roman Catholic church attempted to reinforce its faith through the construction of Neo-Baroque churches while as late as 1880 churches of Protestant denominations in the Eastern Townships tended to adopt conservative Gothic Revival forms rather than the aggressive volumes of the High Victorian mode.[29] An unusually high number of "townships" churches appeared in the CIHB slate roofing printout, but photo verification revealed slate work was invariably plain and uniform in colour. Quebec churches with polychrome slate roofs were designed by Canadian architects of British origin. The zigzags and variously cut slate bands of St. Matthew's Church in Québec (re-designed and rebuilt in the 1870s) and the still surviving diamond slate design of St. George's Anglican Church, Montreal, (Figs. 28 and 29) were the products of William Thomas and his son W.T. Thomas respectively.[30] St. Luke's, Waterloo, Quebec was the conception of Thomas Scott while the 1874 Wesleyan Methodist Church in Aylmer (Fig. 30) with its bordered bands of x's was designed by Langley, Langley and Burke.[31] In the Maritimes, regionally atypical patterned slate roofs of unknown provenance, appeared on St. Dunstan's Basilica in Charlottetown and the still beautifully intact purple and green slate roof of Park Lodge in Halifax (Fig. 31*).[32]

Polychrome-patterned slate roofing was also part of the Second Empire style which shared the Canadian building stage with the High Victorian Gothic in the 1860s and 1870s. Named for its association with the court of Emperor Napoleon III of France, the Second Empire style featured three-dimensional composition using mansard roofs, pavilion massing and rich sculptural detailing.[33] Variously cut and coloured slating was part of the style's repertoire as was carved key stones, pilasters and columns and picturesque roof elements like iron cresting, dormers and clocks. Together, they created the plastic vitality for which the style was so famous. In post-Confederation Canada imitation of this expensive urbanistic mode led to great emphasis on decorative slating. It was as if slate, now increasingly available by rail, could give mansarded buildings lacking the complex massing and sumptuously carved detail of the prototypes, some pretention to Second Empire grandeur.

A review of the Canadian federal buildings built in the Second Empire mode from 1871 to 1881 shows the tendency to polychrome slating in its plainer examples. An important exception was the Toronto Custom House (1873-76), seen by architectural historians as probably the most pompous example of the Second Empire style erected by the Department of Public Works.[34] While it was slated with two colours of plain and round cut slates worked into diamonds between the dormers (Fig. 32), the polychrome pattern is dominated by the elaborately carved stonework of the main façades. Most fine examples of the style with very broken or complex rooflines such as the Montreal Post Office [1872] (Fig. 33), Toronto Post Office (1871-74), and Saint John Custom House (1872-74) used plain or common slate work on the slopes and in the latter two, tin for the domes. In contrast, unusually austere Second Empire designs were dressed up by slate floral motifs or banding in two or more colours that appeared in the towers as well as the slopes (Figs. 34 and 35). In Ottawa such colour banding on the slate mansards of the Post Office (1872-76) and Supreme Court (converted from workshop 1881) [Figs. 36 and 37] was part of a conscious effort to make these buildings accord with the High Victorian Gothic structures on Parliament Hill.[35]

It was in the residential expression of the Second Empire that the full potential of slate for coloured picturesque effect was realized. The more fanciful roofs were probably inspired less by federal architecture than by the pattern books that circulated across the country. The French roofed (mansarded) villas or suburban residences that appeared in *Hobb's Architecture* of 1873 invariably called for slates in two or more tints cut to ornamental shapes.[36] An advertisement for a slate design in Cummings and Miller's *Architecture* [1865] (Fig. 38) revealed that some American architectural firms actually patented specific patterns, but a review of patents here shows there was no Canadian parallel to this practice.[37] Good slaters could emulate most designs and imitation of the examples illustrated in *Bicknell's Village Builder* (1871)[38] made their appearance on Canadian mansards (Figs. 39 to 41).

The majority of polychrome slate mansards that adorned Canadian Second Empire homes displayed one-of-a-kind patterns using common motifs assembled in unique ways. Round forms took precedence over the jagged geometric patterns of the Gothic. The floral motif used on the east and west blocks of Parliament Hill was one of the most popular. It appeared as a simple line of rosettes on round or hexagonally cut slatework (Figs. 42 and 43), flowers within contrasting colour bands (Fig. 44), or the central motif in an intricate design framed by patterned borders (Fig. 41). In one instance (Fig. 45), colour banding was worked into a refined lace-like pattern which echoed the delicacy of iron cresting.

Picturesque, polychromatic slate designs in Canadian domestic architecture were most often identified with the Second Empire mode and yet variously cut slates of uniform colour were the norm for Canadian mansards. The predominant use of all black slating was a natural outcome of supply. The chapter on slate sources indicates that until 1900, 80 percent of the Canadian market was provided from the black slate belt of the Eastern Townships. Government House in Toronto (1868-70) roofed with slate from Quebec's Melbourne, quarry, developed in the 1860s,[39] was an example of a high-style Second Empire residence using continuous black slating (Fig. 46). This manner of applying slate was in greater evidence in multiple housing, schools and religious institutions — buildings that had adopted the mansard idiom as a practical way of providing additional living space. Heavy concentrations of slating have been recorded on mansarded row housing throughout Montreal on such streets as Cherrier and St. Hubert (Figs. 47 and 48). Maison Mère Villa Maria (1878),[40] Collège du Vieux Montréal (1888), and other mansarded Roman Catholic convents and colleges across Canada again illustrated the widespread use of black slate roofing.[41]

The High Victorian principle of polychrome which characterized the slate roofs of so many Gothic Revival and Second Empire buildings in Canada endured from the late 1850s to the late 1880s. Anchored in English re-interpretations of the Gothic Revival and a return to continental and particularly Italian sources, the colourful slate patterns which emulated the geometric mosaics of medieval brick and marble were seen at their jagged Gothic best in the churches of Ontario. The colour banding of variously cut slate equally typical of High Victorian Gothic was continued in the Second Empire mode, where love of the curved form produced some of the most delightfully picturesque, polychrome slate designs in Canadian domestic architecture. Polychrome slating also crept into other styles, such as the 1886-87 renovations of the Neoclassic 1852-53 Brantford County Court House (Fig. 49). Architectural taste, however, ultimately tired of the rich slate designs that were dubbed oil-cloth or calico patterns.[42] The undercurrent of uniformly coloured slating that marked the diffused Second Empire mode grew stronger in the late 1870s and early 1880s.

Single colour slate with scaling in the Queen Anne Revival

The Queen Anne Revival mode led the return to single colour slate work. Queen Anne style had a little to do with Queen Anne and much to do with many other countries and monarchs. The style as it evolved under English architect Richard Norman Shaw and others, showed great admiration for the forms of late 17th- and early 18th-century British building with its mix of classical and medieval features and its ornamental treatment of brick, tile and other materials. The characteristic motifs which came to define the style such as small-paned sash windows, rubbed (decoratively shaped) and cut brick-work, and tile or half-timbered gables were seen as typically English, but it also included shaped Flemish gables, Venetian windows and Japanese-inspired interiors. The point was to blend diverse elements in an original manner to create something comfortable, delicate and graceful.[43] The practitioners of the Queen Anne style loved rich colour but they shunned the strident Gothic combinations in favour of harmonious tones. They usually expressed tones subtly by layering materials such as a red-brick ground storey with tile-hung walls, half-timbered gables and a slate or tile roof. Loud-patterned slate roofs would merely have destroyed the natural refined impression the style was intended to convey.

The Queen Anne architects' attention to colour also produced definite ideas regarding suitable colours for roofs. Red, the colour of the tile which was the characteristic covering for English houses, was the preferred roof colour of the English originators of the mode. J.J. Stevenson, one of the few theorists of the Queen Anne style, endorsed red tile because it was "better in colour than most slates." In his two-volume work on *House Architecture* (1880), Stevenson discouraged the use of Welsh slates, most of which were a "bad" purple colour but recommended Westmorland green slates as "charming in colour, being a pale sea green, which goes well with red brick or against the sky ...."[44] A decade later the colour of Welsh slates was still in disrepute when William Morris, the giant of the Arts and Crafts Movement, strongly criticized "Thin Welsh Blue Slates" as akin to corrugated iron and zinc.[45]

North American architects designing in the Queen Anne mode from the mid-1870s shared the English preference for one-colour red roofs. The shift in focus appeared in the pattern books as early as 1878. *Modern Dwellings*, based on a series of designs Henry Hudson Holly had published in *Harper's Magazine*, noted, "Red for roofs seems to be growing much in favor." It added that introduction of several colours was "objectionable, as it is apt to destroy the repose, and appear frivolous."[46] Unlike their English counterparts, architects in the United States and Canada had access to the red slates of the New York/Vermont slate belt as well as black, grey, variegated, fading and unfading green slates. Because of its proven experience and its availability in a wide range of colour, slate was always more accepted than tile for roofing Queen Anne buildings on this side of the Atlantic. Many designs for the famous American Shingle Style, a Shavian (Norman Shaw) inspired mode named for its indigenous exterior cladding, also called for slate roofs but one colour was *de rigueur.*[47]

By December 1884 the change in slating taste was receiving notice in *Carpentry and Building:*

> *Colour effects are not nearly so popular with present architectural styles as they were formerly ... The tendency [now] seems to be to use a single colour of slate, or at most two shades of the same general colour, and to obtain the principal effect by judicious combination of patterns.*[48]

Canadian slated buildings of Queen Anne Revival influence, the largest stylistic grouping with slate roofs recorded by CIHB, reflect the one-colour roofs typical of the mode. Colour verification of a selection of black and white inventory photos indicated the predominant use of black slate. This finding is consistent with the fact that during this period the Canadian slate industry, which produced blue-black slate, was at its height. In the 1889 Montreal residence commissioned by Senator George Drummond, President of the New Rockland Slate Company, the steep black slate roof with its many turrets and gables was combined with rock-faced red sandstone walls (Fig. 50). As a rule, however, the black slate-roofed Queen Anne buildings of Canada were brick houses situated in Ontario. Slating typically capped a building of brick walls trimmed with stone and tiled or decoratively shingled gables (Figs. 51-56).[49] The polychrome slate roof represented the antithesis of the subtle layering of materials and colours the Queen Anne mode was to convey. It, nevertheless, appeared on a Queen Anne-inspired Châteauguay, residence in one of the most picturesque, if stylistically atypical, slate roofs of this study (Fig. 57*).

Red slate roofs emerged on Canadian Queen Anne Revival buildings of Richardsonian Romanesque influence. It is a moot point whether such buildings should be labelled Queen Anne or Richardsonian. Architect H.H. Richardson was in the forefront of the aesthetic movement in the United States and created the first Shavian manor in American materials. He increasingly drew inspiration from French forms and boldly employed stone in the development of a highly personal style identified by pyramidal massing, and arcaded quarry-faced masonry walls.[50] The features of Richardson Romanesque became the decorative clichés for public building all over North America and in the domestic sphere were liberally mixed with the Queen Anne mode. Many of these hybrid residences were slated black or grey (Figs. 58-60) but red became a favourite of those built with red or buff sandstone (Fig. 61).

Variously cut slate to heighten roof decoration was strongly discouraged by the promoters of Queen Anne and just as strongly ignored by its mass interpreters. The pretty effect of tiles pointed to form a sawtooth line or rounded like fish scales was admitted by J.J. Stevenson, but he added "our

[English] architecture lately has trusted too much for its effect to such little prettinesses ... The glory of a plain tile roof," he noted "is that its mass of quiet beautiful colour gives dignity to the worst and most fantastic architecture."[51] The noble bearing expected of a roof in this period was echoed by H.H. Holly's admonition that "(Tiles) when worked into fancy forms lose their dignity."[52] Nonetheless, fancy forms in slating persisted in Queen Anne architecture in Canada. Like Second Empire slate roofs, however, Queen Anne slatework was nearly always in one colour.

Plain or common slating in the Romanesque Revival, Château and other eclectic modes

With the institutional adoption of the Richardsonian mode in the late 1880s and 1890s, public building in Canada saw a gradual return to continuous slating which was the builder's term for straight-cut slatework uninterrupted by diverse colours or shapes. Discreet bands of scalloped slating appeared on the New Rockland blue slate roof of Redpath Library at McGill University (Fig. 62). The slated roofs of other Richardsonian-inspired buildings such as the provincial legislatures of Ontario (Fig. 63) and British Columbia (Fig. 64), the 1899 Court House in St. John's, Newfoundland (Fig. 65*), and the London Public Library [1894], (Fig. 66), displayed smooth surfaces of continuous or plain work. Taken collectively these buildings also suggest that from the 1890s the black slate roof was being replaced by a wider colour palette that included solid green, red and purple roofs. In the example of the London Public Library red slate again appeared with Scottish red sandstone.[53]

Plain or common slating at its showiest was manifested in the Château style that had its sources in the 16th-century châteaux of the Loire Valley. Though the stately homes of the French nobility were distinguished by their ample roofs sheathed with gleaming surfaces of small slate, the Canadian adaptation for its railway hotels did not often use slating. When it was employed, however, as in architect Bruce Price's hotel station at Place Viger in Montreal, thin, small straight-cut slates expressed the same fine smooth elegance of the French châteaux (Fig. 67).[54] At Craigdarroch Castle in Victoria, (Fig. 68*) an aristocratic residence of mixed Romanesque and Scottish Baronial influences, lustrous planes of red slate highlighted the château roof and provided a vivid textural contrast with the rock-faced stonework of the façade.[55]

The return to symbols exploited by the Château style railway hotels was one of the themes of the Beaux-Arts movement which gained ground in Canadian cities around the turn of the century. Architects, tired of the dabbling in styles which had grown out of Queen Anne, returned to other historical models. In their search for order and clarity, many increasingly went to classical sources. The Chicago World's Fair of 1893 was an ode to monumental classical forms that epitomized the academic reaction. Beaux-Art classicism diminished the roof in the composition of public architecture and this process was completed by the concurrent development of the skyscraper.[56] Symbolic, eclectic architecture in other modes, such as collegiate gothic for schools, would perpetuate slating in the public sector but it was domestic building that set new stylistic trends in slate roofing in the 20th century.

Thick, textured slates in 20th-century domestic architecture

Until the end of the First World War the slate-roofed, architect-designed homes featured in the Canadian professional press, specifically, the *Canadian Architect and Builder*, and *Construction*, exhibited the common slating of the 1890s with increasing use of smaller slates and a wide acceptance of the colour green. Stylistic labels were spurned by the architects of the period but a number of identifiable influences were at play including: English classical, stemming from the full-fledged Colonial Revival in the United States; Tudor Revival, originating from the half-timbering of the 'Queen Anne', and the highly original work of American Frank Lloyd Wright and his English contemporary Charles Voysey, that was firmly anchored in the Art and Crafts Movement's sensitivity to man-made materials and simple building.[57] A feeling of symmetry pervaded much of the new domestic work and translated into roofing this meant, "The playful roof of the 1880s was replaced by a standard hipped or gable one."[58] Responding to this change the slate industry reported an increased demand for the smaller size of slate[59] that magnified the scale of the roof without detracting from the design. The Toronto home of Hugh S. Stevens (Fig. 69), conceived by John Lyle, reflected the new trend to simplicity that relied for its effect on the "poise of the lines and the colour treatment of the walls and roof." Its truncated hipped roof covered with grey-green Vermont slate crowned the façade comprised of red brick lower walls and greyish white stucco above.[60] Similar compositions of green slate roofs with vari-coloured brick walls and half-timbered gables were evident on Montreal houses designed by Nobbs and Hyde, and Phillip Turner, respectively (Figs. 70 and 71).[61]

From 1908, the graduated slate roof received notice in the building periodicals as a way of beautifying slate roofs "marked off into distressingly regular squares or lozenges." The idea was an Arts and Crafts one extolled in the American magazine, *The Craftsman*, and first brought to the attention of Canadians in an article on "Attractiveness in Slate Roofs" that appeared in *The*

Contract Record.[62] The graduated slate roof was actually a reversion to methods of laying slate in the days before sizes or thicknesses were standardized. After sufficient slate was extracted from the quarry the slates were sorted according to length and thickness, then laid with the broader heavier slates close to the eaves and the smaller, lighter slates toward the ridge. The rough texture and uneven edges of the variously sized slates imparted a definite charm to the roof that was enhanced by the use of different colours of slate that often occurred naturally in the same quarry.[63] As *The Contract Record* pointed out, a graduated slate roof could produce the effects of an old roof immediately:

> *Earth and rock colours, combined with the rough surface and ruffed edges of the slates, make it possible to have a roof that from the very first has all the appearance of age and that harmonizes not only with the building but with all the surrounding landscape, because both colour and surface are those of the natural rock.*[64]

It was not only effect but also respect for the inherent qualities of the materials of building that encouraged the renewal of the graduated slate roof. In a lecture on "the Art of Building," M.H. Baillie Scott, a leader with Charles Voysey in the promotion of the artistic house, urged the study of materials as the proper beginning of architecture: "If only instead of ignoring the qualities of materials and forcing them into meaningless forms, we were to begin at the other end, what a new world of art would be disclosed to us." Scott considered the most important quality of roofing materials was texture. "Nothing," he remarked, "is so fatal to the beauty of a roof as tiles which are absolutely regular giving the effect of a surface ruled with absolutely rigid horizontal lines. You might as well cover your building with galvanized iron at once."[65] Scott's lecture, which was published in *Construction* in April 1910, advocated the use of the roughest and thickest slates but criticized slates generally for their failure to yield "to nature's inimitable colouring."[66] This conclusion betrayed his experience of English slate and his ignorance of the gamut of fading and variegated slates in America.

Curiously, despite its discussion and publicity, nearly a decade passed before the graduated slate roof appeared in Canada. The regular notices of domestic work in *Construction* first mentioned graduated slating in 1917 in a description of a "Georgian" red brick residence designed for L.C. Webster in Westmount, Quebec, which it stated, was covered with "rough green slates laid in graduating courses."[67] By 1919 two Toronto homes featured in the same magazine indicated varying colours as well as sizes were being used. The roof of a grey brick residence of classical English influence at Lonsdale and Dunvegan Roads was slated with a mixture of greys, greens and browns including both the unfading and the weathered varieties. Slates of random widths and rough edges were laid with a free graduation toward the ridge. A similar roof of unspecified colours was also used to cover a green shuttered, white stucco home in Cedarvale.[68]

During the 1920s the American slate industry responded to architects' demands for "the appearance of age" in roofs by assembling colour packages of slate designed to project a natural weathered look. Specifications of a Montreal residence being constructed on Drummond Street in 1929, for example, called for slate in a "number 12 colour combination sold by the Roofers Supply Company of Montreal."[69] Tone-on-tone colour mixing in slate roofing continued to appear especially in Canadian domestic work into the 1930s and, as figures 72*, 73*, and 74* indicate, may still be observed in the wealthier residential districts such as Redpath Crescent in Montreal.

Another feature of the heavy or thick slate in this period was its antique reproduction to imitate the appearance of old stone.[70] At the quarry, electrically driven trimming machines cut one- or two-inch thick slate with wavy or irregular edges to emphasize the rustic effect.[71] Such "architectural grades" were marketed under various names, such as "Olde English Rough Cleft Slate" produced by the F.C. Sheldon Slate Company of Granville, New York, and used in the 1927 Gregor Barclay residence of Redpath Crescent, Montreal (Fig. 75).[72] "Tudor Slate" and "Tudor Stone Slate" were other names for rough slates that appeared in such Canadian examples as the fieldstone house of H.P. Edwards, Toronto (1927) and the Tudoresque Montreal Badminton and Squash Club [1928] (Fig. 76*).[73]

Achieving an antique roofing appearance in the 1920s also produced a degree of slate craftsmanship unrivalled in other eras, for old slate roofs implied no evident use of metal flashing. When slate replaced, or actually covered, metal for hips and ridges, it had to be cut to specific angles and applied in various ways to create mitered, Boston or fantail hips (Fig. 77), and saddle or comb ridges. In this period, it was also less acceptable (even if wiser) to have valley metal showing with the result that valleys were closed with slate and sometimes for additional effect, rounded (Fig. 78). If finishing hips, ridges and valleys in slate produced some of the most beautiful slate roofs in Canada, it also created the most expensive. The high cost of slate compared to new synthetic roofing materials ultimately reduced demand and even the luxury trade could not sustain economic volumes of production. The depression of the thirties effectively ended slate as a common roofing material.

With texture as the desired effect, the decorative trends in 20th-century slate roofing ranged from the use of heavy slates about two inches thick, graduated coursing employing slates of diminishing lengths, thicknesses and widths from the eaves to the ridge, colour packages and roughly finished slates

imitating stone. Small slates and a monogram or single-colour roof, particularly green, were favored in the first decade but thereafter larger and heavier
slates with variegated colour effects became popular. The 1920s highlighed
slate roofing craftsmanship not only through its more complicated graduated
slopes, but also in its charming slate hips and valleys which, in special projects,
replaced the almost universal copper and galvanized iron flashings of the
previous century.

III. Canadian Practice in Laying Slate Roofs

The business of slating consisted chiefly in covering the roofs of buildings with slate and all the variations in sizes, laps and fastenings that that could entail. Slate covering was also influenced to an extent by other roof elements, particularly the form, structure, undersheathing and flashing. For any given historical period in Canada the best documentation or most complete description of local slate work is furnished by the extant slate roof. The building historian cannot hope to examine enough roofs to put together a picture of Canadian practice in this way. What I have done, however, is look at the practitioner, the historic influences on the slating trade and the advice dispensed in builders' handbooks and the how-to columns of Canadian architectural journals. Against this background, by sampling building specifications and historic and contemporary photos, it is possible to identify the highlights of Canadian slate roofing practice from mid-19th century to the 1930s.

In 1918, an American periodical published an article entitled "Who does the Slate Roofing in your town?"[1] The question revealed the traditionally close association between the slater and other mechanics in the roofing process. It also showed the blending of roofing trades which was well underway in the heyday of slate roofing in Canada. At mid-19th century, building specifications clearly distinguished between the tasks of the slater, plumber and iron worker on the roof. As the century wore on the functions of each trade were increasingly blurred.[2] Slater's specifications sometimes included lead or copper work and galvanized iron work. "Roofing" specifications described the work of several trades and roofers ads, (Fig. 79) reflected this grouping of expertise. The result was that contracts were awarded to roofing companies that provided all roofing services. It was to take advantage of this situation that the magazine *Metal Worker, Plumber and Steam Fitter,* in 1918, posed the question "Who Does the Slate Roofing...?" Urging his mechanics to learn slating, the editor argued:

> *Tin for valleys, gutters and flashings is needed for slate roofing, and nobody can furnish and place it more advantageously than the tin roofer. This secures favor for his bid for the slate roofing contract. His thorough knowledge of the necessity of making everything wind, weather and waterproof insures care in laying the slate ....* [3]

One of the implications of other roofers entering the slating business was that the importance of experience passed-down from one slater to another was reduced while the technical direction of trade journals and building product suppliers, that is the slate quarries, was correspondingly enhanced. This evolution in the slating trade meant that Canadian slating practice, which started with British and some French influences, ended with almost exclusive dependence on the methodology of its American suppliers.

At the outset of roofing slate production and use in Canada and the United States, both countries showed a strong and natural reliance on British know-how in slate extraction and application. Welsh quarry men initially operated the Quebec, Vermont, New York and Pennsylvania slate quarries. [4] The architects and tradesmen of the 1860-80s period had direct experience of homeland work and these techniques were popularized both by example and the North American publication and distribution of British builders' encyclopedias by Peter Nicholson, Joseph Gwilt, and J.C. Loudon. [5] Canadian and American professional periodicals also printed extracts on slating from the *National Builder* and *Planat's Encyclopédie de l'architecture et de la construction.* [6] By the late 1890s, however, American-generated advice on slating became increasingly evident in the *Canadian Architect and Builder* and was reinforced by the guides and leaflets produced by the American slate industry, the source of most slate used in Canada after 1900.

The pervasiveness of American slating references here, particularly in this century, is unquestionable. American architectural books and construction guides, which were well known in Canada, often contained advice to slaters or sample specifications. [7] The *Canadian Architect and Builder* in answer to inquiries concerning slate drew on American pocketbook authorities such as Kidder and Hodgson. That Hodgson was, in fact, a Canadian could suggest a greater exchange of technical experience between Canada and the United States than his co-authored publications out of Chicago might imply. [8] American dominance of the architectural press in North America was, nevertheless, paramount and was reflected in journals like *The Canadian Engineer* that reprinted articles on slating from American periodicals such as *Stone* and reported American research on slate.[9] The United States quarries and their agents also produced pamphlets for their clients, Canadians included, explaining the kinds of slate and how slates were put on the roof.[10] The National Slate Association, organized in 1922 by American slate producers, architects and contractors published a 1926 handbook, *Slate Roofs*, detailing information on types of slate, roof construction and slate application techniques.[11] Definitions for standard, textural and graduated slate roofs worked out by the Association were the basis of roofing slate standards adopted by the American Society for Testing Materials (ASTM) and influenced slate specifications adopted by the Canadian Bureau of Standards.[12]

The 20th-century standards for roofing slate were building material classifications that did not encompass application practice. In both Canada and the United States, there were never any *official* standards governing how slate was laid on the roof. The present ASTM "Standard Specification for Roofing Slate" (C406-58) covers grades, physical requirements, sizes and thicknesses of slate shingles. Canada has withdrawn this material standard because roofing slate is no longer produced on a commercial scale in this country. Since 1932 the nearest instrument to an official practice standard here and in the United States is *Architectural Graphic Standards*, an illustrated encyclopedia of "proven and current practice" in building construction. Once private, but now published by the American Institute of Architects, this volume has become a standard reference work in every Canadian architects' office. Through the various editions, the portion on slate roofing comprising two tightly illustrated pages on laps, fastenings, hip and ridge techniques remains unchanged.

Long before slate roofing methodology was so visually and concisely summarized in this encyclopedia, professional literature and building specifications pointed to an early Canadian concensus on slating from the requisite associated roof elements of form, structure, sheathing and flashing to the actual practice of laying slate. That this consensus was increasingly American influenced was a reflection of market conditions, the architectural publication industry, the disappearance of the slater's trade as such and roofer's dependence on slate suppliers for information.

Roof form

A steep roof was considered the best roof for slates for both climatic and aesthetic reasons. Lower than a minimum pitch or slope, there was danger that dampness would gradually reach the inside of a joint and damage the construction itself. In a country like Canada where driving rain and melting snow are normal in many cities, steep roofs were particularly important as a precaution against leakage. Height also provided visibility for the slate work. Appearance was often the chief motive for using slate and as the chapter on stylistic trends indicated was integral to many architectural modes.

The minimum pitch for slate roofing most frequently recommended in Canadian building journals was 4-in-12, that is four inches of vertical rise to 12 inches of horizontal run.[13] In historic usage this was usually cited as 4-in-1. A pitch of one third or, still higher, a square pitch (also historic terms) was necessary to achieve a roof profile (Fig. 80). In fact, during the heyday of slate roofing in Canada, slopes usually exceeded the square, in the case of the mansard in particular, approaching the vertical. The Canadian predilection for high pitches for slating was demonstrated early in the 1863 argument that the Quebec jail with its roof slope of 8-in-12, a one-third pitch, was "too flat for slate covering."[14]

The exceptions to the steep rule were the efforts to develop the flat slate roof. In England, James Wyatt, architect to George III, invented a mode of slating with thin large slates screwed to rafter centres, the joints covered with fillets of slate three inches wide, bedded in putty and fastened with screws. Patent slating, as it was known, was said to cover roofs as low sloped as two inches to the foot.[15] In Canada, architect Joseph Scobell of Montreal also patented a similar method of slating designed to cover roofs with pitches of one inch to the foot. Scobell's invention included board undersheathing but also involved a system of fillets covering the joints embedded in a slate cement of his own formula.[16] It is not known whether Wyatt or Scobell's techniques for flat roof slating were ever used in Canada. Elsewhere, Wyatt's system was eventually discredited "from the constant dislodgement of the putty, upon which greatly depended its being impervious to rain."[17]

Advertisements for flat slate roofs[18] appeared in the *Canadian Architect and Builder* from the 1890s (Fig. 79) but the only known instance of such a roof is the National Research Council Building, Ottawa, completed in 1932. This flat concrete slab roof had 3/8-inch-thick black slate laid in a herringbone pattern in mastic over felt and pitch.[19]

Quite apart from the slope, the roof form or shape most suitable to slating also elicited Canadian comment. In 1863 and 1888 architects argued that a roof with a great number of valleys would always retain a certain quantity of snow and ice ultimately forcing water to back up under the slates.[20] Because every gable, dormer and chimney merely added to the problem, suggestions were made to design homes roofed "in a simple and plain manner without unnecessary breaks."[21] The large proportion of gabled, chimneyed and dormered slate roofs which appeared in CIHB's inventory of 3000 extant slated buildings indicates the rule of "no breaks" was widely ignored. An increase in roof pitch may have compensated for the prized irregularities in roofline.

The roof form thus associated with slating in Canada was a square or higher pitch although practice permitted slopes of 1/6 pitch or four inches to the foot. Isolated experiments were made with slating flat roofs but the idea had no apparent popular appeal. The showiness and decorative possibilities of slate meant it was extensively used on the cut-up, picturesque roof despite general Canadian agreement that, in this country, slate would perform best on the unbroken roof surface.

Roof structure

While slate weighed more than other roof coverings such as tin or wood shingles, heavier roof construction was seldom considered necessary to support it until the advent of thick slates in the 20th century. Many of the standard authorities on slate like Nicholson, Gwilt and Planat made no recommendations on roof construction,[22] but pointed out the adaptability of slate covering to either wood or metal framing. With the claims of competing roofing materials in the 1890s the slate industry itself affirmed that normal framing was sufficient to support regular slate, that is slate 3/16-inch thick. "Any building strong enough for a shingle roof and properly called a sage building," explained the Bangor Excelsior Slate Company, "is strong enough for a slate roof."[23] To prove its point the industry argued that the greatest factors in the calculation of the size of timbers necessary to sustain a roof were the allowances for wind, rain and snow pressure or the live load. Since requirements for the live load were the same whatever the material the differential in the dead weight of specific coverings was minimized to the extent that good engineering practice for tin shingles or iron would also accommodate slate.[24]

Heavy slate, from 1/4-inch to two-inch thick, required roof construction designed for the increased weight of the slate as well as walls of sufficient strength to receive the roof load. Heavy slate roofs in Canada and elsewhere were accordingly associated with steel frame or masonry buildings.[25] Depending on the size and weight of the slate, the adjustments in roof construction varied from the choice of wood and spacing, and size of rafters to the tying and bracing of structural members.

Roof sheathing

Historically slate was laid on boards or narrow battens. "When intended to be laid in the best manner," slates were always placed on boarding.[26] Boarding was invariably used on public buildings but the cost reduction by placing battens on open rafters apparently made lathing a popular sheathing in Britain and to a lesser extent in the United States, into the 20th century. The American Bullock (1853) and British Gwilt (1867, 1903) referred to slates nailed to laths or battens as "the common method of slating."[27] *Roofing Slate*, an 1896 publication of a Bangor, Pennsylvania, slate company acknowledged the use of battens for slating in some regions "on account of expense." As late as 1926, industry literature recognized the use of roof lath in many localities.[28] Fir battens were called for in the slating of the Centre Block of the Parliament Buildings, although historic photos and plans fail to indicate if open battening or board sheathing was used.[29] The East (Fig. 81) and West Blocks were board sheathed for slating. A sampling of 30 specifications for Canadian slate roofs suggests such wood under sheathing may have been near universal here.

Board sheathing, which was usually installed by the carpenter, had to meet several conditions as a slating surface. The boards, like those illustrated in the Toronto Customs House (Fig. 83), were to be narrow and well-seasoned to prevent warping or shrinkage and displacement of the slate. The width of roof boards is only occasionally noted in the carpenter's specifications, those of an 1852 Halifax barracks were "not to exceed twelve inches wide," while the boarding of an 1892 Toronto church was not to be more than eight-inches wide.[30] The slate industry recommended six-inch to 10-inch boarding.[31] Boards also had to be even in thickness so that the slate would lie smooth. In Canada, boarding was usually about 7/8-inch or one-inch thick. Best practice demanded matched, planed boards. Nailed to the rafters or tongue and grooved, the boards were placed horizontally although there were instances of diagonal sheathing. Since the 1900s, slate roofs have been double boarded with air space between for insulation.[32]

From the 1850s, the board undersheathing of Canadian slate roofs was covered with tarred roofing felt. The purpose of the felt (Fig. 82) was to protect the roof while the slates were being laid and to form a cushion for the slate. Felt also added insulation value but, despite contemporary claims, added little to the water tightness of the roof.[33] When slate of standard thickness was used one or two ply felt was specified equally; in this century distinctions were made in the grades of felt for heavy as opposed to standard slate.[34] English practice recommended that slates be nailed to battens laid over the felt to create airspace and avoid dry-rot, but the Canadian specifications studied imply that here slates were nailed directly against the felt.[35]

While boarding and felt sheathing were ubiquitous for slating in Canada, 20th-century fireproof construction developments saw the use of concrete and gypsum slabs as undersheathing for slate. The National Research Council building (Ottawa, 1932) provided an example of concrete slab undersheathing for flat slating while several pitched roofs at McMaster University (Hamilton,

1930) had graduated grey and green slates installed over an undersheathing of gypsum slab on steel.[36]

Sheathing, while usual in Canada, was not necessary for laying slate. Slates could be fixed to wood laths on the rafters or attached to a metal, iron or steel framing system. In the former method, the slater himself installed the laths and adjusted them to fit the length of the slates before nailing.[37] Slates were secured to the metal frame by hooks or wires. The French were the leaders in the latter technology and their methods for slating the new Sorbonne and other metal-framed roofs were discussed in American and Canadian periodicals.[38] No Canadian examples of slating over metal have yet come to light.

Flashings

Metal sheets or flashings were placed at the intersections of two roof surfaces or where the roof met a vertical surface such as a chimney or wall. Flashings were designed to prevent water penetration and facilitate drainage, and were installed before or with the slating. Base flashings consisted of metal under the slate and turned up on the vertical surface (Fig. 84). They ran approximately four inches under the slate and eight inches or more up the vertical wall. Cap or counter flashings (Fig. 85*) were started on the vertical surface and bent down over the base flashings. The roof valleys or inside angles formed by two inclined sides of the roof were also lined with metal. The generous valley flashings for Canadian slate roofs (Fig. 86) ranged from 12 inches to 24 inches wide with about 15 inches the most popular.

The materials commonly used for flashing slate in Canada were milled sheet lead, galvanized iron sheeting and copper. Lead was superior for valley and hip work, according to the *Canadian Architect and Builder*.[39] The quality of materials was defined by the ounce weight per square foot. Types frequently specified were Morewood's best No. 26 gauge galvanized iron, lead five or six pounds to the foot and 16 oz. copper.[40]

Laying slate

Once the roof form, structure, sheathing and flashing were determined and far enough advanced, the work of the slate roofer began. Slating started at the bottom or eave with each slate in the row being securely fastened and each subsequent course overlapping and breaking joint (Fig. 87) with the preceding course.

The eave course was always double. In early British practice this was done by nailing a course of larger slates then pushing smaller pieces under the breaking joints.[41] A formula was eventually worked out for determining the length of the short or undereave course.[42] In Canadian and American practice, the short slates were nailed to the eave line then the first course of regularly sized slates nailed over these and properly bonded with the ends of both courses flush (see aa, Fig. 87). A feather-edge lath or tilting fillet was commonly nailed to the eave to start the coursing and keep the subsequent rows tight.

The principle and process of slating hinged entirely upon the lap. The slater's lap was the part or length of a slate overlaying the slate two courses below (see c to d, Fig. 87). At each lap point then there were three layers of slate. The roof slope determined the size of the lap. Historically three inches was the customary lap for the average roof (average meaning a slope from eight inches to 20 inches per foot). Flatter roofs required a four-inch lap and steeper roofs a smaller, or two-inch lap.[43] The Canadian specifications reviewed indicate the three-inch lap was the norm even for the steep mansard. Slight lap variations within a slope were also used such as four inches at the bottom to three inches at the top.[44] In rare instances such as the low-sloped hipped roof of Province House (Charlottetown 1842-47), the lap was as wide as five inches.[45]

The exposure of the slate or the part of the slate to the weather, also known as the gauge, (see b to c, Fig. 87) was found by deducting the lap (three inches) from the length of the slate used and dividing by two.[46] According to this formula the 24-inch-long Duchess slate which was popular in Canadian public buildings of the mid-19th century would have shown about 10-1/2 inches to the weather. From the 1870s, the Federal Department of Public Works specified Countess size slate, with 8-1/2 inches exposed. The gauge varied on the graduated slate roof, that of the W.R.G. Holt residence in Montreal was reduced from seven inches at the eaves to five inches at the ridges.[47] As a general trend, slate exposures or gauges in this country saw a gradual reduction from the 1850s to the 1930s although the gamut of slate sizes remained available to the client.

The common method of securing slates in Canada was by nailing with two nails near the head. For slate of normal thickness the holes were made by the roofer or slater himself with a slater's hammer or punching machine available from the 1860s. Heavy slates were drilled at the quarry. The nail holes, as illustrated by slates removed from Laurier House and Earnscliffe in Ottawa, were located about 1-1/2 inches from the outside edges and about a quarter of the slate length from the upper edge. The types of nails generally specified were copper or galvanized iron, the latter the most popular. The length of the nails varied from 1-1/2 inch to two inch with 1-3/4-inch nails the standard for federal buildings. An interesting exception to nailing was the two "swedes iron screw bolts" called for slates of the 1879 Custom House in Saint John, New Brunswick.[48]

In addition to nailing, slates were sometimes bedded in hair mortar or from the 1870s in elastic patent cement. Rendering was originally associated with the British and Irish practice of slating on battens. Mortar with hair to make it stick to the laths was intended to keep the slates from rattling, particularly in windy places near the seacoast.[49] In Canada slates were set in hair mortar at Province House (Charlottetown, 1842-47).[50] A Canadian building magazine was still advocating "torching" in 1899 "to keep out draughts and prevent leakage on exposed sites."[51] About 15 percent was added to the roofing cost by laying slate in elastic cement according to the *Canadian Contractor's Handbook of 1901*.[52] This extra expense undoubtedly limited use of cement to vulnerable hip and ridge areas and the spaces created by slates of different thicknesses on special effects roofs. Experience had also probably proven that the silencing and wind guard provided by cement was unnecessary on the boarded, felted undersheathing characteristic of slate roofs in Canada.

To finish laying the slate, special methods were developed for cutting and fastening slate at intersecting surfaces such as hips and ridges. "Comb" and "saddle" ridges and "saddle" or "mitered" hips were traditional old country methods of providing slate apexes for the slate roof. From the mid-19th century, slate roofs were also finished with ornamental slate roll ridge pieces, consisting of a seven-inch wide wing with a 2-1/2-inch roll worked on the top edge.[53] The evidence of historic photos suggests Canadian roofers generally avoided these slating techniques by finishing the roof under iron and copper cornice rolls or terracotta ridge pieces (see figures throughout the text and particularly 65*, 88* and 89*). The wide use of iron cresting in both Gothic Revival and Second Empire styles promoted the galvanized iron ridge on the slate roof, but even the non-crested roofs of the Queen Anne Revival were commonly metal ridged. Twentieth-century residences such as the 1928 Fetherstonhaugh home in Montreal (Fig. 73*) provide the few known examples of slated ridges in Canada. Slates hips, also known as closed hips, were documented more often. Prior to its finishing with slate, the hip was lined with sheet lead or galvanized iron, then slating was continued over the metal, brought up to the edge of the hip by fantail, miter or other cuts (Fig. 77) and bedded in elastic cement for added security. The saddle hip which consisted of slate laid on a cant strip over the hip joint and covering the hip slates appeared in homes of the 1910-20 period.[54] Despite these variations, hips of copper or galvanized iron coping dressed onto the slate appear to have been the norm in Canadian slate roofing practice.

Slating had to be accommodated to roof valleys as well as hips and ridges. Valleys, as indicated, were always metal lined or flashed and slates were lapped over this metal on either side. In an open valley the slates did not come together leaving the metal flashing exposed. A closed valley meant slates completely covered the metal. The open valley (Fig. 90) was typical in Canada, the slates trimmed to produce a space two inches wider at the bottom than at the top to allow for snow sliding.[55] The closed valley, less practicable in the Canadian climate but more aesthetically pleasing, was identified in the cable house at Heart's Content (Fig. 91), Newfoundland and as previously discussed in the domestic architecture of the 1920s (Figs. 72*, 74*, 78). A variation of the closed valley type was the round valley illustrated in the Montreal Badminton and Squash Club (Fig. 78). Establishing the contour for this work was part of the wood sheathing contract. Slates for the round valley had to be trimmed to the proper radius and required such careful workmanship that the technique was seldom used.[56]

Slate roofing in Canada today is usually associated with restoration work. To supplement the foregoing historic overview of Canadian slating practice, a number of publications provide detail on specific procedures. The succinct *Architectural Graphic Standards*, already mentioned, capsulizes the more detailed *Slate Roofs*, a 1926 publication of the U.S. National Slate Association, recently reproduced by and available from the Vermont Structural Slate Company. *Slate Roofs* has become the reference bible on slating practice including flashing, repairs, roof construction and sample specifications. Useful how-to articles on roofing with slate, reflecting renewed interest in the covering, also appear in the popular *Old House Journal* (May 1980) and *Fine Home Building* (April, May 1984). The latter article, although not applicable to most Canadian slate roofs, provides a clear step-by-step description of installing slate over open roof lath. *Ardoise*, a French journal devoted to contemporary slate roofing, is also available in Canada. Finally, Restoration Services, Canadian Parks Service, has prepared a master specification for slate roofing which is available upon request.

IV. Sources of Roofing Slate, Past and Present

When Canadians used slate for roofing they obtained supplies of the material from two principal sources, the Eastern Townships of Quebec and the eastern United States (Table 5 and Figs. 92, 95 and 96). Slates from these areas have been ascribed to similar geological formations, mostly the Cambrian system, but they differ widely in colour, hardness and durability.[1] Identifying the origin of slate on extant slate roofs can therefore be problematic. Surviving specifications and contracts usually document slate but by names that are often puzzling. I had to study the slate industry itself to identify the major producing regions, the characteristics that distinguished the slates of each area and the names by which they were marketed. With this information it was then possible to suggest the probable origin of most slate in Canada and sources for matches of historic slate.

Canadian sources

The centre of the Canadian slate industry was Quebec's Eastern Townships. Through southern Quebec three ridges of rocks extend in a north-easterly direction to form the Sutton Mountain, Stoke Mountain and Lake Megantic hills. The basins between these ranges are lined with rocks of Cambrian age at their borders and by Ordovician strata in the centre. Where these systems contact serpentine and other intrusive rocks, slate deposits occur. Historically, the deposits of greatest commercial value were located in the county of Richmond along the Melbourne slate band (Fig. 92), a belt about a quarter-mile wide running southwest and northeast from Melbourne township across the St. Francis River into the townships of Cleveland and Shipton. Within a

15-mile distance four quarries were opened: the Danville quarry in Shipton, range IV, lot 7; the Steele or Bedard quarry in Cleveland, range XV, lot 6; the Melbourne or Walton quarry (Fig. 93), Melbourne, range VI, lot 22, and the New Rockland quarry, Melbourne, range IV, lot 23. The Walton and New Rockland quarries were the principal sources of roofing slate production in Canada.[2]

Melbourne and Richmond slates, the terms used in specifications to designate slate from the two main quarries, were similar in character. An 1876 *Catalogue of the Economic Minerals of Canada* described them as a "bluish-black colour, smooth surfaced, thin, light and strong."[3] Their composition according to analysis of the Geological Survey of Canada strongly resembled slate from Angers, France which was used on seminary buildings in Montreal and withstood the effects of that climate for over 100 years.[4] Quarry operators compared New Rockland slate to American slate (probably the coarser Vermont coloured slate) using the analogy of hard wood to soft wood.[5] The hardness or strength of the Canadian slate made it costly to work and put Canadian producers at a disadvantage relative to some, although certainly not all, American quarries.

Coloured slates were never successfully commercially quarried in Canada. A green slate was taken from Jervis Inlet, British Columbia for local work in the 1890s.[6] In the early 1900s, before Newfoundland was part of Canada, quarries at Smith's Sound, Trinity Bay, produced large quantities of purple and some green slate for English markets.[7] This slate was also used for roofing buildings throughout the Avalon peninsula, and as the first chapter indicated, showed up in inventory data.

Coloured slate quarries operated intermittently in the Eastern Townships. Montreal builder Joseph Scobell opened a purple slate quarry at Kingsey, range I, lot 4, in the 1850s but the enterprise was abandoned for lack of capital within two years.[8] The Rankin Hill Slate Quarry, on lot 25, range V in the township of Acton, produced red and green slate "for ornamental work in slate roofing" between 1875 and 1877 before low prices and the lack of market contributed to its failure.[9] As in Newfoundland, coloured slate was taken from deposits in southern Quebec for local roofing needs.[10]

During the 1880s and the 1890s the blue-black slate of the Melbourne band supplied an average 80 percent of the Canadian market (Table 3, Chapter 1). After a lack of capital to modernize equipment forced the Walton quarry to close in 1878,[11] the New Rockland quarry was the sole producer. One of their most important clients was the federal government which slated nearly every post office and custom house it built in eastern Canada in the seventies and eighties. According to mining returns filed by quarry owners with the Geological Survey of Canada, the principal markets of the New Rockland Company were in Toronto, Montreal and "the lower provinces."[12] During the 1880s the monopoly which the Company enjoyed in Canada enabled it to limit the sale of roofing slate to four or five firms in Toronto, thereby keeping up the price, a practice strongly condemned by the *Canadian Architect and Builder*.[13] By 1914 the output of the New Rockland Company was entirely handled by the Roofers Supply Company of Toronto.[14] However, by then the New Rockland production was less than 10 percent of Canadian consumption (see Table 4 in Chapter 1).

A change in the tariff which had protected the Canadian slate industry since 1881 was one of several factors in the Canadian shift to imported slate after 1900. Between 1881 and 1897 a duty of 80 cents per square was charged on blue or black foreign slate and $1 per square for red, green or other colours.[15] The higher duty on coloured slate was intended to discourage the importation of inferior grades such as fading sea green slate which was plentiful and cheap in nearby Vermont.[16] In 1897 the Laurier government imposed an *ad valorem* duty of 25 percent on all roofing slate, not to exceed 75 cents per square.[17] The abolition of the distinction between coloured and blue or black slate cleared the way for the admission of all grades of slate into the Canadian market. The lowering of protection came at a time when recession had already reduced the manpower and production of the New Rockland quarry and Canadian architects had begun to show antipathy for "the monotony of black slate roofs."[18] As taste changed toward greater texture and colour for roofs, shingles came into favor as well as coloured slates which were by nature thicker and rougher. Not only was the United States the nearest source of coloured slate but, because domestic black slate output was so drastically reduced, after 1900 American quarries became Canada's principal suppliers of roofing slate.

American sources

The phenomenal growth of the American roofing slate industry after 1893 helps explain the difficulties of the New Rockland Company and Canadian post-1900 reliance on American sources. Even at the height of the Canadian industry in 1888 United States production was 30 times the value of Canada's, by 1906 the multiple had reached 180.[19] In terms of quantities, during the first decade of the 20th century the American industry was producing about 250 times our domestic output. U.S. totals amounted to about 1 250 000 squares in contrast to 5000 squares for Canada's.[20] Exports accounted for five to 15 percent of the value of American roofing slate production. Two-thirds of the exports went to the United Kingdom and about 12 percent each to Canada and

British Australia.[21] Canada's consumption of roofing slate was a miniscule proportion of the annual American production and yet American slate provided on average 75 percent of Canadian requirements from 1900 to 1910. (For comparison, see Table 4 in Chapter 1 and Table 5 here.)

American slate (current samples, Fig. 94) coming into Canada before 1900 was mainly of the blue-black variety with the exception of the years 1882 to 1887 when red, green and other colours led the imports (Table 6).[22] Coloured slate imports undoubtedly increased after 1900, but it is questionable if they surpassed blue-black as its volume of production remained far greater. Four states commercially quarried blue-black roofing slate. The origin of the Canadian imports is not clear from Canadian and American trade statistics. Price lists of building materials that appeared in the *Canadian Architect and Builder* in 1888 and 1889 specify three types or trade names of black slate, indicating two states as probable Canadian suppliers. The slates listed were "black Lehigh" and "black Chapman's" both from Pennsylvania and "unfading black Monson" from Maine.[23]

Pennsylvania was always the largest American producer of slate, its output constituting about 60 percent of the total national production from the 1880s to 1914.[24] Commercial slate was obtained from three regions: 1) the Peach bottom area in Lancaster County and the southeastern part of York County extending into Maryland, 2) Lehigh County and 3) Northampton County (Fig. 95). Two distinct belts or types of slate crossed the latter counties. All the slate quarries of Lehigh County and those in the northern part of Northampton County belonged to the soft vein belt which consisted of fading black carbonaceous slates. These slates had a high grade of fissility, were fairly sonorous and had some magnetite. The lower or southerly belt known as hard vein occurred in Northampton County only. Slate there was of a dark-grey shade, fine texture, slightly rough and slightly lustrous. It belonged to the fading series but discoloured less rapidly than slate of the soft vein.[25]

"Black Lehigh" slate as listed by the *Canadian Architect and Builder* probably referred to slate from the Lehigh region, an area of three-square miles comprising about 50 quarries centred around Slatington. The Bangor region of Northampton County outranked Lehigh in commercial importance, however, the Lehigh region had a greater volume of export trade.[26] A highly ferrous carbonate content was an objectionable feature of the slates of Slatington for its oxidation contributed to discolouration, but their fine cleavage and ease of quarrying made them reasonable to supply, an important consideration when duty was involved.[27]

"Black Chapman's" slate originated in the hard-vein belt at Chapman, Northampton County. The principal quarry there was operated by the Chapman Slate Company. The slate was dark bluish-grey although slightly lighter in colour than the soft-veined Lehigh slate. Because of its hardness, Chapman slate like the Canadian New Rockland, had holes drilled at the quarry whereas the softer slates were simply holed by the slater himself.[28]

Unfading black Monson slate was produced in the centre of the state of Maine, in the southern part of Piscataquis County. Monson slate may have referred specifically to slate taken from quarries near the village of Mason or, more generally, to slates of the Monson district which also included quarries at Blanchard and Brownville. The chief roofing slate quarries at Monson itself, the Monson Pond quarry and the Matthews quarry, produced a dark bluish-grey graphitic mica slate, rough and with little lustre. At Brownville, a very dark grey "black" slate distinguished for its fine texture, smooth and bright cleavage, high grade of fissility and sonorousness was assessed by Dale as "one of the best slates for roofing in the United States."[29] Maine ranked third in American slate production in 1898 yet it produced just six percent of the national total.[30] Exports of Maine slate to Canada would likely have been correspondingly small in comparison to quantities from Pennsylvania (Fig. 96).

At least two other "brands" of American blue-black slate must have reached the Canadian market at one period or another although they are not mentioned in the price lists discovered. These were "Bangor" and "Virginia" slate. "Bangor" a word obviously borrowed from Wales, was the market name for dark grey slate from the Bangor-Pen Argyll region in Northampton County, Pennsylvania, the largest producing area in the United States. Slate from Pen Argyll and Albion, a quarry in the Pen Argyll region, was sold under these names as well as included in the larger designation of "Bangor." "Virginia" blue-black slate was a lustrous granular slate quarried in Albemarle and Buckingham counties, Virginia.[31]

Red, purple, unfading green and green slates were also listed in the building materials priced by Canadian builders' periodicals in 1888, 1889 and 1890.[32] Although the origin of these slates is not stated, studies of the American slate industry reveal that the chief source of American coloured slate (Fig. 97) imported into Canada had to be the Vermont/New York slate belt. This belt produced about 28 percent of American slate, ranked second to Pennsylvania in national production[33] and marketed the greatest volume of coloured roofing slate in the United States. Red roofing slate, in fact, was never produced commercially in any other state but New York. Albemarle County, Virginia quarried dark green and variegated (green and purple) roofing slate but volume of output was small and the area farther from Canada than the Vermont/New York region.[34] Slate from the Vermont quarries, the most cheaply produced in the United States, was seen by Canadians themselves as the chief challenge to their slate industry (Fig. 97). "One of the reasons why Canadian production has been small," the *Canadian Architect and Builder* editorialized in June 1899, "is because slate can be imported very cheaply from the Vermont region in the United States, notwithstanding the duty."[35]

The Vermont/New York slate district (Fig. 97) extended north-south along the New York, Vermont boundary between Washington County on the west and Rutland County on the east. All but one of the quarries in Ordovician red slate were on the New York side and most of the quarries in Cambrian greenish and purplish slate were on the Vermont side. Five principal colours of slate were marketed from the region. New York red slate, quarried in Granville and Hampton, was a lustreless, rough slate, reddish brown in colour, that became brighter instead of fading on exposure. Because of its occurrence in thin and irregular deposits, red slate was costly to quarry and the most expensive of American slates. Unfading green slate came from the northern part of Rutland County, Vermont. It was a greenish-grey slate with a slightly irregular texture and lusterless surface. Sea green slate quarried from Poultney in Rutland County to West Rupert in Bennington County, Vermont, varied from a light grey to greenish grey when freshly quarried but after a few years of exposure due to the oxidation of iron carbonate, changed its colour to brownish grey. The fissility and tendency to discolour of sea green slate made it the cheapest of American slates. "Purple," a dark purplish brown slate alternated with the greens and was either fading or unfading according to the green with which it was interbedded. "Variegated" or "mottled" slate was greenish grey and purplish mixed.[36]

American output of coloured roofing slate increased in the early decades of the 20th century but never exceeded the production of blue-black slate.[37] The output of specific colours fluctuated at difference periods according to the industry reports that appeared in the annual publication, *Mineral Resources of the United States*. Sea green slate more than doubled any other colour in quantity and value in the pre-war years. During 1909, in Rutland County the centre for the Vermont slate industry, 65 sea green quarries were at work in comparison to 25 producing other colours.[38] Mottled or variegated slates, then purple slates were the next in total value while unfading green slate, the highest grade of green slate, was in much less demand.[39] Production of New York red slate, smallest in comparison to the other colours, reached its largest number of squares between 1908 and 1912.[40] In the post-war years the output of variegated and unfading green slate increased and that of sea green declined.[41]

If American roofing slate production reflected American consumer demand, it did not necessarily suggest Canadian consumption patterns. With the 1897 abolition of separate duties for coloured and uncoloured slates, however, there are no import statistics relating Canadian's 20th-century use of coloured slate to blue-black. An increased use of coloured slate in Canada after 1900 is indicated by architects' impatience with black slate, the popularity of textured, coloured roofs,[42] and the increasing number of grey-green, green, red and variegated slate roofs described in building magazines. The comparison of American blue-black and coloured slate production is nevertheless interesting since output determined the choice of slates available to Canadian architects and builders whether or not it directly influenced or reflected their taste.

In summary, the American roofing slate industry was largely confined to five states of the eastern United States — Pennsylvania, Vermont, Maine, New York and Virginia. Pennsylvania and Vermont consistently quarried almost 90 percent of the American product during the years the United States was the major supplier for the Canadian market.[43] Slate from each of these states was distinct in colour and quality. The principal product from Vermont was coloured, — green, purple, mottled or variegated — while the Pennsylvania slates ranged from dark grey to bluish-black. Vermont/New York slate was always prized for specialized work and gained acceptance in the 20th century for ordinary roofing. Nevertheless production of Pennsylvania dark grey so outdistanced coloured slate in volume that it is probable its use continued to be standard for roofing in this century.

Guidelines for determining slate origin

The small number of roofing slate quarries in North America and the distinct character and production of each suggest certain signposts for the identification of extant slate roofs in Canada.

Prior to 1900 blue-black slate on roofs constructed here would generally have come from the Melbourne slate band in the Eastern Townships since this region provided more than 80 percent of the Canadian market then. From 1878 black Canadian slate was produced exclusively by the New Rockland Slate Company of New Rockland, Quebec. During the 19th century, small amounts of black slate were also imported into Canada from the states of Pennsylvania and Maine, with the largest amount going to Ontario (Table 7).[44]

After 1900, most blue-black slate used in Canada was American in origin. The states of Pennsylvania, Maine and Virginia were the probable suppliers with Pennsylvania far in the lead. Pennsylvania slate was marketed under such names as Bangor, Pen Argyll, Albion, Lehigh, Chapman's and Peach Bottom. Generally speaking these slates were dark grey in colour and all fading with the exception of the Peach Bottom. Blue-black slates from Maine known as "Monson" and from Virginia, called Buckingham or Virginia, belonged to the unfading series. The Virginia slates were perhaps the most lustrous although this was a quality that varied from deposit to deposit within states.

Coloured slate employed in Canada was usually American in origin with the exception of regions in Newfoundland and southern Quebec where coloured slate was available and quarried for local roofing needs. Red slate came exclusively from New York state, although it was sometimes marketed by Vermont quarries. Fading sea green, unfading green, purple and variegated slate were mainly the products of Vermont and to a very small extent of New York.

Present roofing slate sources

The historic slate roofing quarries in the Eastern Townships have ceased operating. This means Canada's main source of roofing slate during that vigorous period of its use — the blue-black type covering so many Canadian public buildings — is no longer available. However, in Newfoundland, the Island Tile and Slate Company is producing green and purple-red roofing slate. Opened in 1986 from an old quarry in Nutcove, Trinity Bay, the Company supplies markets in eastern Canada and the United States. It is the only roofing slate quarry currently in commercial production in Canada.

Four of the original five main producing states of the American roofing slate industry still have quarries in operation. Blue-black slate is available from the Buckingham-Virginia Slate Corporation, Richmond, Virginia, and the Structural Slate Company in Pen Argyle, Pennsylvania. Vermont companies also produced a black slate. Red slate, still quarried exclusively in Granville, New York is provided by two New York companies, Evergreen Slate Company and Hilltop Slate. These companies and two Vermont quarries, Rising & Nelson and Vermont Structural Slate also produce the full range of colours historically quarried in the Vermont/New York region, including unfading mottled green and purple, unfading green, unfading purple, and weathering green. Names and addresses of companies are provided in the endnote.[45]

With the gamut of colours still supplied by the United States roofing slate industry (Fig. 94), it is nearly always possible to find replacement slate for the Canadian slate roof. A clean section of the underside of the old slate can be compared to samples from various American quarries, or alternately, sent to the slate companies for analysis. More than one suitable colour-fit can often be found. For example, blue-black New Rockland Canadian slate has recently been matched to Buckingham-Virginia slate on Laurier House, Ottawa and replaced by Pen Argyl, Pennsylvania slate on Windsor Station, Montreal.

The durability of roofing slates can vary. The ASTM Standard Specification for Roofing Slate[46] has established three grades, based on the length of service that may be expected: Grade S1 — 75 to 100 years; Grade S2 — 40 to 75 years, and Grade S3 — 20 to 40 years. Each grade designation must meet certain physical requirements for rupture, water absorption and weather resistance.[47] The user specifies the grade, requests the test results from the quarry and should retest the slate when delivered to ensure it meets the specified requirements. Roofing slate durability is not necessarily related to geographic origin. The prevailing opinion on longevity, according to the *Old House Journal*, is that "Pennsylvania slate lasts at least 50 years, Vermont (and N.Y.) slate lasts at least 100 years, and Buckingham (VA) slate lasts at least 175 years."[48] Another view suggests Pennsylvania slate lasts "at least 75 years."[49] Since grades can vary within a quarry, it is best for the purchaser to ask for the verifying tests of the grade ordered.

Table 5
Squares of Roofing Slate Imported by Country 1882-96, 1912-34*

Year			Country		
	U.K.	Belgium	France	Netherlands	U.S.
1882					653
1883					630 1/2
1884					485
1885					647
1886					326
1887					543
1888					3184
1889					8417
1890					3207
1891	45				7939
1892					15 552
1893					12 816
1894					3067
1895					1470
1896					2891
1912	543	199		39	19 432
1913	1121	370	125		19 841
1914	1846	40			19 816
1915	1506	58			14 830
1916	114				7093
1917	70				4258
1918					3716
1919					8144
1920					4699
1921					6662
1922					6095
1923					7028
1924					5675
1925	27		65		4941
1926					4512
1927					5297
1928					7053
1929	51	109	35		6791
1930	710		61		8170
1931	16				4219
1932					3687
1933					1192
1934	97				1038

*Separate statistics for roofing slate are unavailable from 1896 to 1912.

Compiled from: *Canadian Mineral Statistics 1886-1956; Mining Events 1604-1956*, Reference Paper No. 168 (Ottawa: Queen's Printer, 1957), and *Tables of the Trade and Navigation of the Dominion of Canada* (Canada, Dominion Bureau of Statistics, Industry and Merchandizing Division, General Statement ... Roofing Slate).

Table 6
Squares of Roofing Slate Imported by Colour 1882-97

Year	Blue or black	Red, green or other
1882	305	348
1883	201 1/2	429
1884	51	434
1885	86	561
1886	49	277
1887	94	449
1888	2527	657
1889	7719	698
1890	2339	868
1891	6841	1098
1892	7401	750
1893	6166	6650
1894	2995	72
1895	1180	290
1896	2421	470
1897	1235	331

Tables of the Trade and Navigation of the Dominion of Canada. (Canada, Dominion
Bureau of Statistics, Industry and Merchandising Division, General Statement ...
Roofing Slate).

Table 7
Squares of Imported Slate by Colour and Province 1882-96.
B + b = blue and black; r + g = red and green; o = others.

Year	Colour	Province						
		Ont.	Que.	N.S.	N.B.	Man.	B.C.	P.E.I.
1882	b + b	118	50	3	134			
	r + g, o	254	57		37			
1883	b + b	163		1/2	38			
	r + g	360	53			16		
1884	b + b	51						
	r + g	376	58					
1885	b + b	60						
	r + g	364	187		26			10
1886	b + b	49						
	r + g	208	69					
1887	b + b	86			8			
	r + g	350	98	1				
1888	b + b	2487	20				20	
	r + g	472	185					
1889	b + b	7719						
	r + g	444	127	2			125	
1890	b + b	2186			33	120		
	r + g	591	217				60	
1891	b + b	6334	28		19	260	200	
	r + g	855	214			29		
1892	b + b	7374			27			
	r + g	587	163					
1893	not aportioned among provinces							
1894	b + b	2991	4					
	r + g	60	12					
1895	b + b	1180						
	r + g	139	147	4				
1896	b + b	2421						
	r + g	319	30	25				

Canadian Mineral Statistics 1886-1956; *Mining Events 1604-1956*, Reference Paper No. 168 (Ottawa: Queen's Printer, 1957).

V. Conclusion

In Canada, the use of slates for covering roofs was a late 19th-century urban phenomenon. Documented in the French Regime, slating remained uncommon until railways developed and quarries opened in the 1860s. Slate roofing attained its greatest popularity in eastern Canadian cities in the 1880s when architectural styles featuring decorative slating were in vogue and competition from factory-made materials was still slight. Roofing slate production statis tics and import tables precisely identify 1889 as the high point of Canadian consumption. Slating continued in common use until the First World War, but from that period on appeared almost exclusively on architect-designed buildings and homes, eventually disappearing in the 1930s.

Evaluation of buildings recorded by the Canadian Inventory of Historic Building confirmed the chronological and demographic trends in slating suggested by the historical record. The low western Canada record of slate roofs coincided with regional urban development in the post-1900 period when slate use was generally declining. What became apparent from the Inventory data that was not evident from other sources was the distribution of roofing slate in the East. In the Atlantic region slate roofs were commonly restricted to public and ecclesiastical architecture, although local slate supply in Newfoundland produced many types of slate-roofed buildings there and an inventory total greater than the other three Atlantic provinces combined. In the core slate roofing areas of Ontario and Quebec, the CIHB recorded four slate-roofed buildings in Ontario to one in Quebec. The Quebec examples were mainly in Montreal and villages of the Eastern Townships near the slate quarries. The city of Toronto provided the largest concentrated sampling of slate roofing in the country. Many residences recorded there clearly emphasize the popularity of slate roofing in 19th-century urban Canada.

Stylistic trends in slate roofing were not geographically defined but associated instead with architectural fashion. Slate size, thickness, colour and cut were the main determinants of roof appearance and, in different periods, one or more of these features were emphasized by the arbiters of taste. In this book five distinct decorative trends in slating are identified. The first was ornamental slating in the Gothic Revival popularized by A.J. Downing's advice to residential builders to alleviate the monotony of the steep roof by cutting shingles or slates in diamond, round and hexagon shapes. Slate had barely come into use here at this time and Canadian examples of this phase are exceptional.

In the second decorative trend influenced by the High Victorian Gothic and Second Empire modes, polychrome and patterning were added to ornamental scaling. New interpretations of medieval diapering and marquetry in English church architecture provided the inspiration for the colourful patterns in slate. The zigzags, diamonds and colour banding so typical of this phase appeared in nearly every province. They were best and most numerously illustrated in the High Victorian Gothic churches of Ontario. American design books popularized pattern slate roofs in Gothic and French-roofed or mansarded styles. Some of the most picturesque slate roofs in Canada emerged in Second Empire examples which accented the curved form and floral patterns in contrast to the jagged lines of the Gothic.

A reaction to these crazy "calico" roofs marked the third stylistic trend which saw a return to one-colour slate work led by the Queen Anne Revival. While ornamental scaling was retained in the beloved towers of the Queen Anne Revival style, a uniform colour quickly became ubiquitous. In the CIHB sampling, slate-roofed buildings in this mode represented the largest grouping of historic types. Red slate roofs, preferred here to the red tile endorsed by English Queen Anne architects, most often emerged in Canada on sandstone buildings of mixed Queen Anne-Richardsonian Romanesque influence.

The fourth stylistic trend in slate roofing was marked by plain or common slating unvaried in colour or shape. Many eclectic modes at the turn of the century exemplified this phase but it was best illustrated by the smooth roofs of the Château style. Finally, a taste for texture stimulated by Arts and Crafts Movement's sensitivity to materials prompted the use of thick, rough slates in 20th-century domestic architecture. Occurring when slate was not a common but a luxury roofing material, this fifth decorative trend manifested itself in expensive graduated and textural roofs, tone-on-tone colours and antique effects. The five principal variations in slating trends were thus related to architectural periods, and, in many cases, found their origin in specific styles.

The review of historic Canadian slating techniques from the perspective of written and photographic evidence, made with the proviso that the extant slate roof would always be the best document, showed little change in practice from mid-19th century to the 1930s and indicated a wide concensus on the best methods. Generally Canadian architects and builders slating roofs in this country preferred high pitches (from most photos probably square or higher), board and felt undersheathing, double eave courses, three-inch laps, head nailing with two nails (galvanized iron or copper) 1-1/2 to two-inch long, open valleys wider at the bottom to allow snow sliding, and cement added on windy hip and ridge areas. Metal hips and ridges were standard while slate, in these areas, exceptional. Saddle and comb ridges, closed and sometimes rounded valleys, were usually seen on the expensive domestic architecture of the 20th century. A Canadian patent for a flat slate roof was issued in 1854, but the only such roof identified in this study was the 1930 National Research Council Building, Ottawa. Concrete and gypsum slab undersheathing for slate were used in 20th-century institutional architecture. No Canadian examples of slating over open wooden lath or metal frame were discovered, although observations in the field could indicate the practice was more common than historical research has suggested.

A useful guidepost emerging from this study was the importance of American sources in Canadian slating practice. Many, if not most, of the popular design books, pocket manuals and encyclopedias referencing slating were American. How-to columns in the Canadian professional press were frequently extracted from United States periodicals. Canadian roofing suppliers circulated practice booklets produced by United States slate quarries and the United States National Slate Association's, 1926 technical summary, *Slate Roofs*. In both Canada and the United States, there were never any official standards for laying or installing slate on roofs. From 1932, diagrams of slate roofing methods were illustrated in *Architectural Graphic Standards*, an encyclopedia of current practice in building construction. This volume, published six times since then, and a reprint of *Slate Roofs* (both American) provide the main references used by Canadian architects regarding slating practice today.

The two main source areas for roofing slate employed in Canada, documented here, were the Eastern Townships and the eastern states of the United States. Before 1900 about 80 percent of the Canadian market was supplied from the blue-black Melbourne slate band in the county of Richmond, Quebec. For many years the New Rockland Slate Quarry was the sole producer. Recession in the 1890s and an 1897 change in the Canadian tariff reduced domestic output so that after 1900 about 75 percent of Canadian roofing slate was American in origin. Quarries from Pennsylvania, Maine and New York/Vermont were the main suppliers. Red slate was the exclusive product of New York, most other colours originated in Vermont and the blacks, blue-blacks and greys came from Pennsylvania, Maine and Virginia.

The blue-black slate of the New Rockland quarries which adorned so much of Canada's public and private architecture of the 1870-1900 period, and which in numerous buildings still survives, is no longer being produced. Green and purple slate are currently produced in Trinity Bay, Newfoundland. Some quarries remain open in each of the eastern United States slate regions, except Maine. Matches for almost any colour of slate can therefore be obtained using the information provided in the text. American slate is graded for durability and users should request the grade suited to the length of service required.

In urban Canada today, the recycling of commercial and government buildings and the revitalization of housing in older neighbourhoods have renewed interest in slate and other historic building materials. Many slate-roofed buildings endure from the 1870s to 1930s period. This stock of slate roofs reflects the standardization of slating practice influenced by American technical literature and industry advice. But the decorative treatment of slate is definitely varied. The stylistic trends identified in this book provide restoration guidance for specific examples. As urban heritage preservation continues, similar historical perspectives on wood, stone and metals may be pivotal in ensuring the variety and historical integrity of our built past.

Illustrations
Colour and Black and White

12* Yates Castle, Brantford, Ontario

0* Wellington Street United
Church, London, Ontario, 1878

25* The Necropolis (with roof close-up), Toronto, 1880-85

26* St. Peter's Basilica (with roof close-up),
    London, Ontario, 1880-85

31 Park Lodge, Halifax, Nova Scotia

57* Laporte Residence, Pointe St. Louis, Châteauguay, Quebec

65* Court House (left), St. John's, Newfoundland, 1899

68* Craigdarroch Castle, Victoria, British Columbia, 1885-90

72* 1414 Redpath Crescent, Montreal, 1929

73* 1291 Redpath Crescent, Montreal, 1928. Architect of both houses:
H.L. Fetherstonhaugh

74* 1300 Redpath Crescent, Montreal

76* Tudor Stone Slate." Montreal Badminton and Squash Club,
1926-27. 3505 Atwater Ave., Montreal

85* Base flashing above eyebrow window and cap or counter flashing below. Note also rusting galvanized iron ridge pieces

88* Galvanized iron flashing, Wellington Street United Church, London, Ontario, 1878

89* Problems with a terracotta tile ridge, Park Lodge, Halifax

94* Colour samples of American slate still available

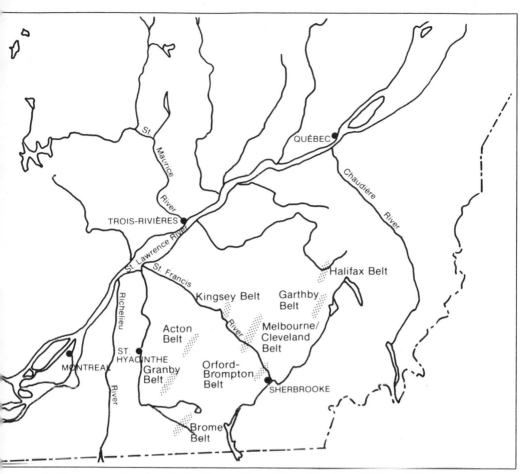

1 Slate quarry locations in the Eastern Townships of Quebec, 1850s and 1860s.
(*Base map from Mines Branch, Energy, Mines and Resources Canada,
Adapted by Janet Weatherston*)

Walton Slate Quarries.

These Slate Quarries are now fully opened, and the undersigned is prepared to supply a very Superior **Roofing Slate** of a beautiful blue color, and of a quality not inferior to the best **Welsh** Slate; or any other known. The **Quarries** are situated near the G. T. R. R., in the Township of Melbourne, C. E.

The following is a list of **Prices** for 1861, shipped on the Cars at Richmond; which together with the low rate of freight, as per annexed Schedule, will enable parties to have Slating done at, or under the **London** Prices.

Sizes.	No. in a Square.	Price per Square.		Sizes.	No. in a Square.	Price per Square.
24 x 14	98	$4.00		16 x 10	222	$3.60
24 x 12	114	4.00		16 x 9	246	3.60
22 x 12	126	4.00		16 x 8	277	3.60
22 x 11	138	4.00		14 x 10	262	3.40
20 x 12	141	3.80		14 x 9	291	3.40
20 x 11	154	3.80		14 x 8	327	3.40
20 x 10	169	3.80		14 x 7	374	3.20
18 x 11	175	3.80		12 x 8	400	3.00
18 x 10	192	3.80		12 x 7	457	2.75
18 x 9	213	3.80		12 x 6	533	2.50

Rate of freights from Richmond, Canada East, per Grand Trunk Rail Road.

TO	Per Car.	Per Ton.	Per Square.	TO	Per Car.	Per Ton.	Per Square.
POINT LEVI.	$20.00	$2.00	$0.50	BELLEVILLE...	$37.50	$3.75	$0.94
MONTREAL...	17.50	1.75	0.44	COBOURG.........	37.50	3.75	0.94
PRESCOTT ...	27.50	2.75	0.69	TORONTO........	40.00	4.00	1.00
OTTAWA	40.00	4.00	1.00	GUELPH	50.00	5.00	1.25
BROCKVILLE	30.00	3.00	0.75	LONDON..........	50.00	5.00	1.25
KINGSTON	32.50	3 25	0.82	SARNIA...........	60.00	6.00	1.50

Orders addressed to the undersigned, at Melbourne, Canada East, or Toronto, Canada West, will be promptly attened to.

BENJAMIN WALTON,
Proprietor.

Toronto, March 21, 1861.

2 Circular sent to Federal Department of Public Works, 1861, announcing opening of a Canadian slate quarry, RG11, Vol. 372, no. 52201. (*National Archives, C-125885*)

3 Centre Block, Parliament Buildings, Ottawa, ca. 1880. (*National Archives, C-10006*)

4 University College, Toronto, ca. 1891. (*National Archives, RD-382*)

5 Ottawa River elevation of Parliament Buildings showing striped slate roof.
(*National Archives, PA-10663*)

6 Some slated buildings in Newfoundland: Carbonear (top), Trinity, Petley

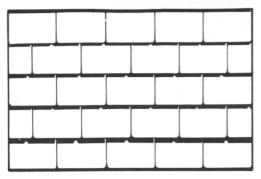

1) Plain or common

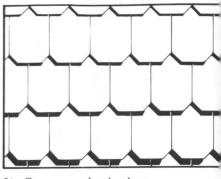

2) Corner angles broken

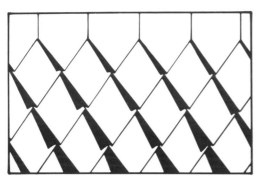

3) Lozenge or diamond form

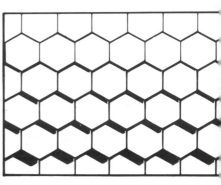

4) Hexagon cut

5) Round cut

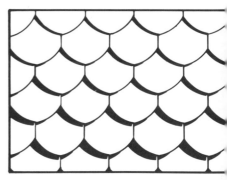

6) American cottage or, with a sharp point, Gothic.

7 Six basic slate shapes or cuts. (*Illustration by Wayne Duford*)

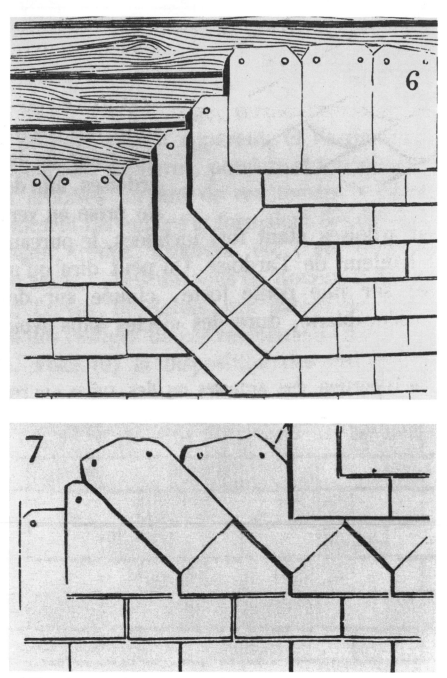

8 The quincunx (top) and the saw tooth patterns as illustrated in E. Viollet-le-Duc,
   Dictionnaire raisonné *de l'architecture française du XIe au XVIe siècle*,
   pp. 455, 456. (*Adapted by Wayne Duford*)

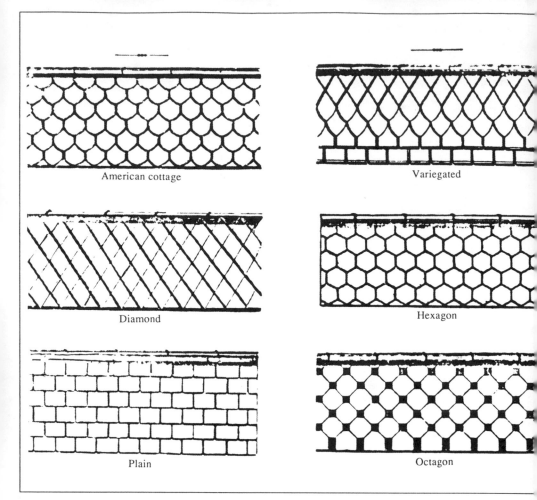

American cottage

Variegated

Diamond

Hexagon

Plain

Octagon

9 Patterns of ornamental slate for roofing available from the Eagle Slate Company in Vermont 1857 lustrated in *The Ready Calculator*, John F. Trow, New York, 1857. (*Adapted by Wayne Duford*)

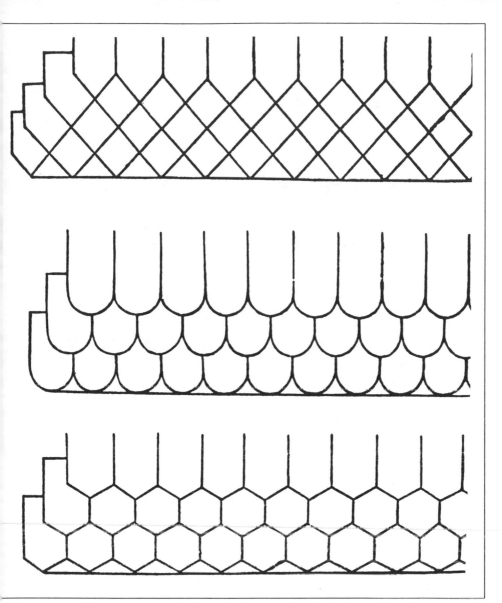

10 Slate and shingle shapes or cuts recommended by A.J. Downing in *Architecture of Country Houses*, 1850, p. 181. (*Adapted by Wayne Duford*)

11 192 Durham, Madoc, Ontario.

13 West Elevation of University College, Toronto, showing slate banding, zig-zags and diamonds. (*National Archives, RD-582*)

14 All Saint's, Margaret Street, London, England (1850-53), by William Butterfield.
(Photo from George W. Hersey, High Victoria Gothic, Johns Hopkins Univer-
sity Press, Baltimore & London, 1972, p. *109)*

15 Centre Block Tower from slate roof of West Block, Ottawa, ca. 1873-78. (*National Archives, C-7*

16 Bijker residence, Wallacetown, Ontario

17 Residence, Newboro, Ontario

18 Slate patterns in Toronto, St. Michael's (1845-66) and its neighbour, the Metropolitan Methodist Church (foreground), 1872. (*National Archives, PA-32093*)

19 Central Presbyterian Church, Cambridge, Ontario, 1880

21 Metropolitan Methodist Church, Toronto, ca. 1890s

22 Welland Avenue United Church, St. Catherines, Ontario, 1877

23 Dorchester Presbyterian Church, Dorchester, Ontario, 1889

24 St. George's Anglican Church, London, Ontario, 1890

27 St. Patrick's Church, Hamilton. (*National Archives, PA-32652*)

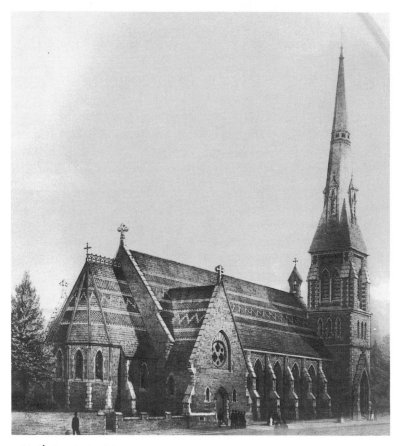

28 Église Saint-Mathieu, Quebec City, (*National Archives, PA-46025*)

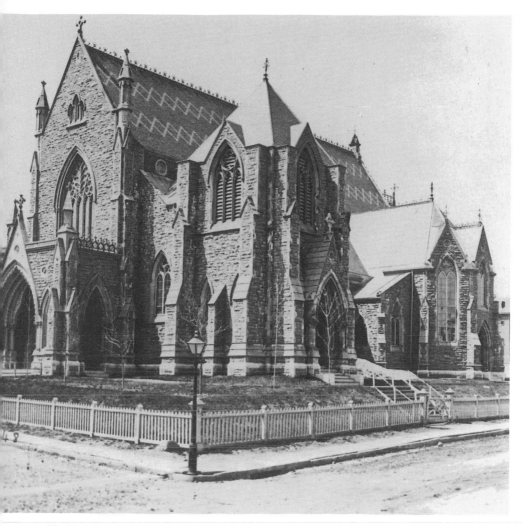

29 St. George's Anglican Church, Montreal. (*Notman Photographic Archives, McCord Museum, Montreal, No. 84, 748-I*)

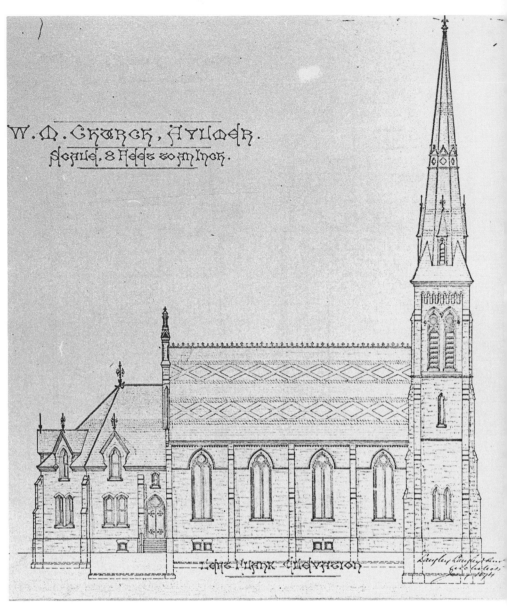

30 Architect's design for Wesleyan Methodist Church, Aylmer, Quebec, 1874.
(*Ontario Archives, L-1036*)

32 Toronto Custom House (1873-76), demolished 1919. (*National Archives, PA-46479*)

33 Montreal Post Office (1872-73), demolished. (*National Archives, PA-53140*)

34 Post Office, Custom House, Inland Revenue, Guelph, Ontario (1876-78), demolished. (*National Archives, PA-46462*)

35 MacKenzie Building, Royal Military College, Kingston (1876-78). (*National Archives, PA-16837*)

36 Ottawa Post Office (1872-76), demolished 1938. (*National Archives, PA-51843*)

37 Supreme Court, Ottawa, converted from work shed, 1881. (*National Archives, PA-8389*)

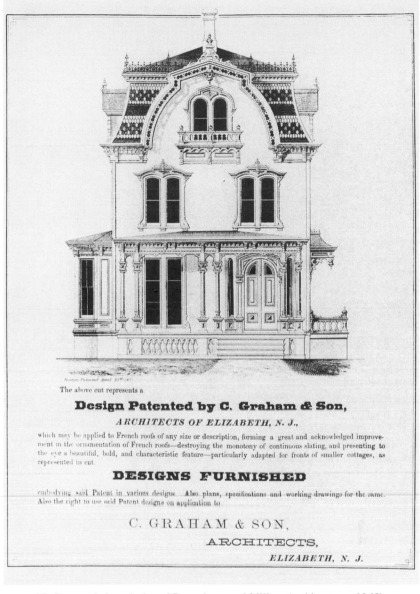

The above cut represents a

Design Patented by C. Graham & Son,

ARCHITECTS OF ELIZABETH, N. J.,

which may be applied to French roofs of any size or description, forming a great and acknowledged improvement in the ornamentation of French roofs—destroying the monotony of continuous slating, and presenting to the eye a beautiful, bold, and characteristic feature—particularly adapted for fronts of smaller cottages, as represented in cut.

DESIGNS FURNISHED

embodying said Patent in various designs. Also plans, specifications and working drawings for the same. Also the right to use said Patent designs on application to

C. GRAHAM & SON,

ARCHITECTS,

ELIZABETH, N. J.

38 Patented slate design. (*Cummings and Miller, Architecture, 1865*)

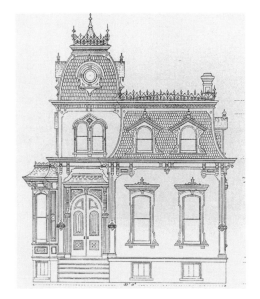

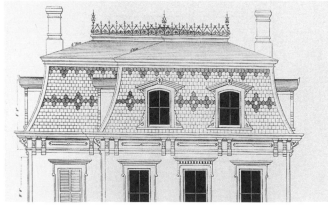

39 Mansard roof slate design. (*From Bicknell's Village Builder, 1871*)

40 Mansard slate designs: 168 Jackson Street, Hamilton, (top)
and 165 Moore St., Sherbrooke, Quebec

41 Floral mansard slate design, 67 Queen St., Guelph, Ontario

42 445 Talbot St., Aylmer West, Ontario

43 49 Ridout St. S., London, Ontario

44 3 Meredith Crescent, Rosedale, Ontario, 1876

45 Col. Lazier's House, Belleville, Ontario. (*National Archives, PA-10059*)

46 Government House, Toronto (1868-70). (*National Archives, RD-541*)

47 Slate row, Cherrier St., Montreal

48 Slate row, St. Hubert St., Montreal

49 Brant County Court House, Brantford, Ontario (1852-53);
polychrome slate roof added 1886

50 Hon. George Drummond's Residence, Montreal, 1889. (*Notman Photographic
Archives, McCord Museum, Montreal, No. 2458 View*)

Miss Julia Greenshield's residence, 2 Elmsley Place, Toronto, 1897. (*National Archives, PA-123379*)

52 "Waverley," 10 Grand Ave., London, Ontario, 1877

53 29 West St., Brantford, Ontario

54 417-19 Drew St., Woodstock, Ontario, 1886

55 192 Daly St., Ottawa, 1888

56 Drawing of house for Wm. Goulding, St. George Street, Toronto,
(*Canadian Architect and Builder, January 1891, p. 4*)

58 17-19 Madison Ave., Toronto, 1893

59 18 Madison Avenue, Toronto, 1893

60 1803-1819 Dorchester Blvd. West, Montreal, ca. 1890

61 Residences on Dorchester St. W., Montreal, with red sandstone and red slate roofs, 1890s. (*Canadian Architecture and Builder, June 1892*)

62 Redpath Library, McGill University, Montreal (1890-91). Architects: Sir Andrew Taylor and Gordon. (*Notman Photographic Archives, McCord Museum, No. 11, 310 View*)

63 "Queen's Park," Legislature of Ontario, 1893. Architect: P.B. Waite. (*National Archives, RD-322*

64 Legislative Buildings, Victoria, 1898. Architect: Rattenbury.

66 London Public Library, London, Ontario, 1894, demolished
1954. (*Canadian Architect and Builder*)

67 Place Viger, CPR Hotel and Station, Montreal, (1896-98). (*Notman
Photographic Archives, McCord Museum, Montreal*)

69 Toronto residence of Hugh S. Stevens, 1910. Architect: John Lyle.
(*Construction*, Vol. 3, *No. 8 [July 1910], p. 76*)

70 595 Pine Ave., Montreal, Quebec, built 1911 (green slate). Architects: Nobbs
and Hyde (*Construction*, *Vol. 8, No. 6 [June 1915], p. 269*)

71 W.E. Mowat Residence, 646 Carleton Ave., Westmount, Quebec, 1916. Architects: Phillip Turner & Careless (*Construction, Vol. 9, No. 6 [June, 1916], p. 199*)

75 "Old English Rough Cleft Slate." Barclay Residence, Redpath Crescent, Montreal, 1927. Architect: H.L. Fetherstonhaugh.

77 Slate hips: (from top) fantail, mitered, saddle

78 Slated valleys: closed (top), round.

ROOFERS

79 A sample of roofers ads. (*The Canadian Architect and Builder, 1890s*)

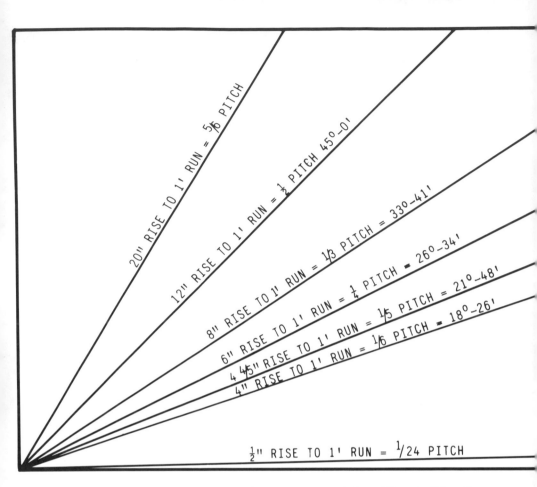

The following labels appear on the radiating lines of the diagram:

20" RISE TO 1' RUN = 5/6 PITCH

12" RISE TO 1' RUN = 1/2 PITCH 45°-0'

8" RISE TO 1' RUN = 1/3 PITCH = 33°-41'

6" RISE TO 1' RUN = 1/4 PITCH = 26°-34'

4 4/5" RISE TO 1' RUN = 1/5 PITCH = 21°-48'

4" RISE TO 1' RUN = 1/6 PITCH = 18°-26'

½" RISE TO 1' RUN = 1/24 PITCH

80 The ideal roof pitch for slate was a square pitch or 45° angle (12 to 1). (*Illustration by Wayne Duford*)

1 East Block construction 1862-64. Note board sheathing on tower. (*National Archives, C-37299*)

82 Felt over board sheathing of West Block, 1860s. (*National Archives, C-7715*)

83 Board sheathing on concave mansard of Toronto Customs House, 1875. (*National Archives, PA-46270*)

84 Base flashing running under slate

86 Open valley with wide flashings, typical
of most Canadian slate roofs

How Slate is Put On.

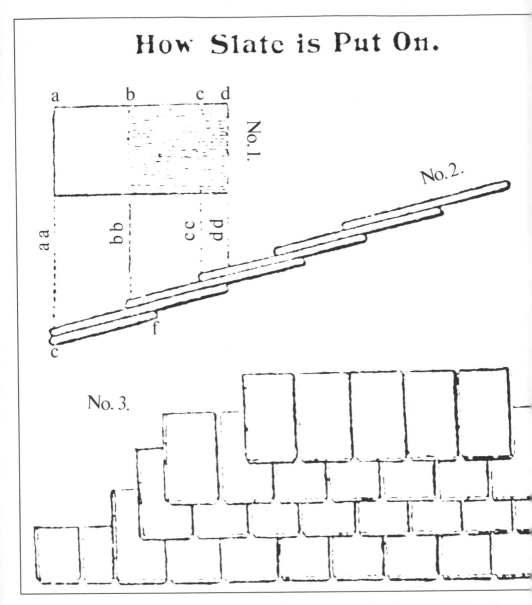

87 Illustration from Bangor Excelsior Slate Company's "Roofing Slate 1896," p. 6

90 Open valley on slate roof of St. Peter's Basilica, London, Ontario (1880-85)

91 Closed valley, Cable house, Heart's Content, Newfoundland

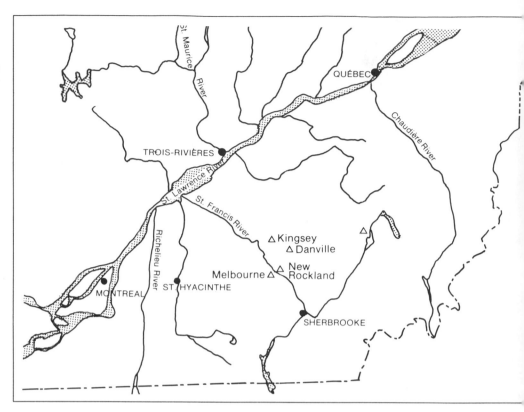

92 The chief Canadian slate quarries, Eastern Townships, Quebec. (Reproduced by Janet Weatherston)

93 Melbourne slate quarry, Eastern Townships, Quebec (*National Archives, PA-40112*)

134

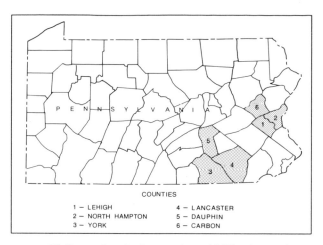

COUNTIES

1 – LEHIGH   4 – LANCASTER
2 – NORTH HAMPTON   5 – DAUPHIN
3 – YORK   6 – CARBON

95 Pennsylvania slate regions. (*J. Weatherston*)

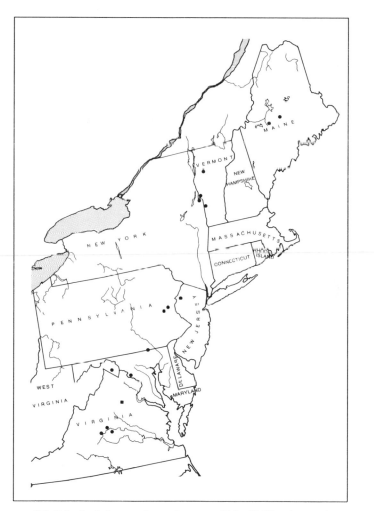

96 Principal slate regions of eastern U.S. (*J. Weatherston*)

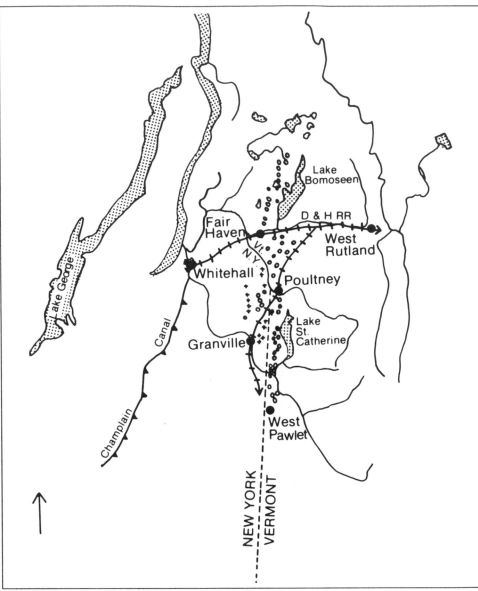

97 New York/Vermont slate belt. (*J. Weatherston*)

Endnotes

I. A Late 19th-century Urban Phenomenon

1 Jean Lindsay, *A History of The North Wales Slate Industry* (London: David and Charles, 1974), pp. 14, 88, 185, 193.

2 T. Nelson Dale, et al., "Slate in the United States," United States, Department of the Interior, *Geological Survey Bulletin No. 586* (Washington: Government Printing Office, 1914) (hereafter cited as T. Nelson Dale, "Slate in the United States"), pp. 19-25; Oliver Bowles, "The Characteristics of Slate," *American Society for Testing Materials, Proceedings of the 26th Annual Meeting*, Part 2, Technical Papers (26-29 June 1923), pp. 525-27.

3 Luc Noppen, *Les églises du Québec (1600-1850)* [Éditeur officiel du Québec/Fides, 1977], pp. 7-8.

4 Luc Noppen, *Notre-Dame de Québec: son architecture et son rayonnement (1647-1922)* [Quebec: Éditions du Pélican, 1974], pp. 23, 27.

5 Olivier Maurault, *La paroisse; histoire de l'église Notre-Dame de Montréal* (Montreal: Thérien Frères, 1957), p. 177.

6 Ramsay Traquair, *The Old Architecture of Quebec* (Toronto: Macmillan, 1947), p. 14.

7 Yves Laframboise, *L'architecture traditionnelle au Québec* (Montreal: Les Édition de l'Homme Ltée, 1975), p. 55. "Les Couvertures en ardoises durent longtemps, ... et les réparations ne sont pas grandes au lieu qu'une couverture en bardeau il faut changer souvent et réparer de même."

8 Pierre Mayrand, "Les matériaux de couverture en Nouvelle-France aux XVIIe et aux XVIIIe siècles," *Bulletin of the Association for Preservation Technology*, (hereafter cited as *APT Bulletin*), Vol. 2, Nos. 1-2 (1970), pp. 71-73.

9 Peter Kalm, *Travels in North America* (New York: Dover Publications, 1966), Vol. 1, p. 430.

10 Blaine Adams, "The Construction and Occupation of the Barracks of the King's Bastion at Louisbourg," *Canadian Historic Sites: Occasional Papers in Archaeology and History* (hereafter cited as *CHS*), No. 18 (1978), pp. 71-72.

11 Peter N. Moogk, *Building a House in New France* (Toronto: McClelland and Stewart, 1977), pp. 56-57.

12 Joseph-Noel Fauteux, *Essai sur l'industrie aux Canada sous le régime français* (Quebec: Ls-A. Proulx, 1927), Vol. 1, pp. 137-60; *Bulletin des Recherches Historiques*, Vol. 16, No. 6 (1910), pp. 167-192.

13 Pierre Mayrand, op. cit., p. 74.

14 Ramsay Traquair, op. cit., p. 14.

15 Pierre Mayrand, op. cit., p. 74. "De toutes les couvertures la meilleure dans ce pays est celle en ardoise qui résiste à la gelée."

16 Elizabeth Simcoe, *Mrs. Simcoe's Diary*, ed. Mary Q. Innis (Toronto: Macmillan, 1965), p. 61; Isaac Weld, Jr., *Travels through the States of North America and the Provinces of Upper and Lower Canada during the years 1795, 1796 and 1797*. Vol. I (London: n.p., 1800), p. 336.

17 Isaac Weld, op. cit., p. 336; John Lambert, *Travels through Canada and the United States of North America in the years 1806, 1807 and 1809*. Vol. I (London: C. Craddock and W. Jay, 1814), pp. 51-52.

18 John Lambert, op. cit., Vol. I, pp. 51-52; Diana S. Waite, "Roofing for Early America," in *Building Early America: Contributions toward the History of a Great Industry*, ed. Charles E. Peterson (Radnor, Penn.: Chilton Book Company, 1976), p. 148, note 35.

19 National Archives of Canada (hereafter cited as NA), MG24, D 11, Vol. I, Visits to Canada and Newfoundland 1808-9, Reports by Jenkin Jones of the Phoenix Assurance Company of London, 25 August 1808, p. 93.

20 NA, WO55, Miscellanea, Vol. 1551(7), North American Provinces, Commissioners' Report, 9 Sept. 1825, Col. Sir James Carmichael Smyth, R.E. to Duke of Wellington (Master General of the Ordnance).

21 Elizabeth Vincent, "The Royal Engineer's Use of Roofing Materials in British North America 1820-1870," Manuscript in progress, Environment Canada, Parks Service, Ottawa; John Joseph Greenough, "The Halifax Citadel 1825-60: A Narrative and Structural History," *CHS*, No. 17 (1977), pp. 73-74; ibid., "South Magazine, Halifax Citadel, A Structural History," Manuscript Report Series (hereafter cited as MRS) No. 223 (Ottawa: Parks Canada, 1977), pp. 38-39.

22 NA, Admiralty Papers, Vol. 504, Oct. 1813, pp. 299-303; Mary K. Cullen, "*A History of the Structure and Use of Province House, Prince*

Edward Island, 1837-1977," MRS No. 211 (Ottawa: Parks Canada, 1977), p. 14 and App. 13; NA, MG13, WO 55, Vol. 863, p. 418, R. Bydon to Gen., Mann 19 June 1826.

23   NA, MG24, D 11, Vol. 2, Visits to Canada and Newfoundland, 1845-46, Reports by John Joseph Broomfield of the Phoenix Assurance Company of London, 17 Oct. 1845, pp. 50-53; ibid., 25 Oct. 1845, pp. 54-57; ibid., 8 Nov. 1845, p. 70.

24   Ibid., 20 Sept. 1845, pp. 18-19.

25   *The Quebec Gazette*, 10 June 1846, n.p.

26   Canada. Department of Public Works, *Annual Report*, 1867, App. No. 23, Report by G.F. Baillairgé, "Description and Cost of the Public Buildings Constructed or Improved by the Department of Public Works," pp. 248-84.

27   George A. Nader, *Cities of Canada* (Canada: Macmillan, 1975), Vol. 1, pp. 167-80.

28   Ibid.; Diana S. Waite, op. cit., p. 140; Philip C. Marshall, "Polychromatic Roofing slate of Vermont and New York," *APT Bulletin*, Vol. 2, No. 3 (1979), p. 77.

29   Mathilde Brosseau, "Gothic Revival in Canadian Architecture," *CHS*, No. 25 (1980), p. 7; Christina Cameron and Janet Wright, "Second Empire Style in Canadian Architecture," *CHS*, No. 24 (1980), pp. 7-14.

30   Oliver Bowles, "The Technology of Slate," United States, Department of the Interior, Bureau of mines, *Bulletin No. 218* (Washington: Government Printing Office, 1922) [hereafter cited as Oliver Bowles, "The Technology of Slate"] p. 20; Philip C. Marshall, op. cit., p. 77.

31   *Tables of the Trade and Navigation of the Province of Canada for the year 1855* (Toronto: Stewart Derbishire and Georges Desbarats, 1856), p. 162; ibid., 1856, p. 101; ibid., 1857, p. 100; ibid., 1858, p. 172; ibid., 1859, p. 160; ibid., 1860, p. 118; ibid., 1861, p. 112; ibid., 1862, p. 115; ibid. 1863, p. 117; ibid., 1864, p. 102; ibid., 1865, p. 108; ibid., 1866, p. 101.

32   Canada. Geological Survey, "Report of Progress for the Year 1847-8," *Journal of the Legislative Assembly of the Province of Canada*, 1849, App. G.

33   Ibid., "Report of Progress for the Year 1849-50," *Journal of the Legislative Assembly of the Province of Canada*, 1850, App. V., "Catalogue of Economic Minerals and Deposits of Canada and Localities"; J.C. Taché, *Descriptive Catalogue of the Productions of Canada Exhibited in Paris in 1855*, (Paris: G.A. Pinard-Dentan and Co., 1855), 14th class.

34   Canada. Geological Survey, *Report of Progress from the Commencement to 1863* (Montreal: Dawson Brothers, 1863), pp. 830-31; Robert Bell,

C.E., *Roofing Slate as a Source of Wealth to Canada: A Visit to the Walton Slate Quarry*, A paper read before the Natural History Society, Melbourne, Canada East, 8 October 1863; NA, RG11, Vol. 372, No. 52201, "Walton Slate Quarries."

35 *Documents Relating to the Construction of the Parliamentary and Departmental Buildings at Ottawa* (Quebec: Department of Public Works, 1862), pp. 70-71.

36 *Descriptive Catalogue of a Collection of the Economic Minerals of Canada and Notes on a Stratigraphical Collection of Rocks. Philadelphia International Exhibition, 1876* (Montreal: Lovell Printing and Publishing, 1876), pp. 106-7.

37 "The Slate Industry of Newfoundland," *The Canadian Architect and Builder* (hereafter cited as *CA&B*) Vol. 15, No. 7 (July 1902), p. 109; ibid., No. 11 (Nov. 1902), pp. 170-71; G.F. Carr, *The Industrial Minerals of Newfoundland*, Mines Branch Publication No. 855 (Ottawa: Department of Mines and Technical Surveys, 1955), pp. 133-37.

38 Christina Cameron and Janet Wright, op. cit., pp. 8, 13-14.

39 *General Report of the Minister of Public Works for the Fifteen Years from 30 June 1867 to 1 July 1882* (Ottawa: MacLean, Roger and Company, 1883), App. No. 2. Thomas Fuller, Chief Architect, "Report on Public Buildings throughout the Dominion," pp. 148-208.

40 Christina Cameron and Janet Wright, op. cit., pp. 15-17.

41 Mathilde Brosseau, op. cit., pp. 20-26.

42 Ibid., p. 22.

43 The breakdown in Table 3 of roofing slate and school slate production in Canada is recorded in NA, RG87, Vol. 34, No. 140, Records of the Minerals Resources Branch, Slate Production Information, 1885-1919.

44 Tables 4 and 5 have been compiled from Canada, Dominion Bureau of Statistics, Industry and Merchandising Division, *Canadian Mineral Statistics 1886-1956; Mining Events 1604-1956*, Reference Paper, No. 168 (Ottawa: Queen's Printer, 1957), p. 64 and *Tables of the Trade and Navigation of the Dominion of Canada*, General Statement by Countries and Provinces of the Total Quantities and Values of Merchandise Imported, item, Roofing Slate, 1886-1936. The latter tables are printed separately until 1889 and subsequently as part of the *Sessional Papers*.

45 Canada, Geological Survey, *Annual Report*, 1888-89, Part K, R.W. Ells, "On the Mineral Resources of the Province of Quebec," p. 130. Ells states the average production at the New Rockland quarry was about 2200 squares of roofing slate a month. The *Canadian Architect and Builder* reported production at 3000 squares per month. See "The Canadian Slate Industry," *CA&B*, Vol. 2, No. 7 (July 1889), p. 82.

46 "The Canadian Slate Industry," *CA&B*, Vol. 2, No. 7 (July 1889), p. 82.

47 "Lead, Copper and Zinc for Roofing," *CA&B*, Vol. 9, No. 12 (Dec. 1896), pp. 202-3.

48 Christopher A. Thomas, "Dominion Architecture: Fuller's Canadian Post Offices, 1881-96," Thesis, University of Toronto, 1978, pp. 95-96.

49 Canada. *Sessional Papers*, 1890, No. 9, p. 39; Canada. House of Commons, *Select Standing Committee on Public Accounts: Reports, Minutes of Evidence and Exhibits in connection with the Langevin Block, 1891* (Ottawa: Queen's Printer, 1891), p. 66.

50 NA, RG87, Vol. 34, 1892-97.

51 See Tables 4 and 5 and Chapter 3.

52 George A. Nader, op. cit., p. 206.

53 In his description of Montreal working class domestic architecture of the period, Marsan states that flat roofs generally replaced mansards. A letter to the *CA&B* in 1904 acknowledges this but criticizes the fact that slate slips were often placed above the cornice. Some mansard slate roofs did continue such as exhibited by the roofs of the residential zone of Viauville in Maisonneuve. See Jean-Claude Marsan, *Montréal en évolution* (Montreal: Fides, 1974), pp. 277-79; "Montreal Letter," *CA&B*, Vol. 17, No. 196 (April 1904), p. 73; Paul-André Linteau, "Town Planning in Maisonneuve," *Canadian Collector*, Vol. 13, No. 1 (Jan./Feb. 1978), p. 83.

54 T. Ritchie, *Canada Builds: 1867-1967* (Ottawa: National Research Council of Canada, 1967), p. 280.

55 Genuine Bangor Slate Company, "Slate and Its Uses," 2nd ed. (Easton, Penn.: 1907), pp. 24-32. This booklet is in the possession of the Historical Society of Pennsylvania.

56 "Prepared Roofing," *The Contract Record*, Vol. 24, No. 31 (Aug. 1911), pp. 48-49.

57 Charles H. Stringer, "Modern Roofing Materials -- Part III: Asbestos Shingles and Slates," *The American Architect*, Vol. 114, No. 2231 (25 Sept. 1918), pp. 62-63.

58 The track channeler was a machine, run by steam or by compressed air which travelled back and forth on a track and cut out a channel in the slate rock. It cut smooth surfaces and reduced the amount of waste obtained by previous methods of blasting.

  The wire saw, also used for making cuts in the quarry or sawing blocks on the bank, consisted of a three-strand steel cable of 3/16-inch or 1/4-inch diameter running as an endless belt which when fed with sand and water sliced the rock. See Oliver Bowles, "The Technology of Slate," pp. 31-33; ibid., "The Wire Saw in Slate Quarrying," United States,

Bureau of Mines, Technical Paper 469 (Washington: Government Printing Office, 1930), pp. 1-31.

59 The marketing problems peculiar to slate are outlined by Oliver Bowles, "Consumption Trends in the Roofing-Slate Industry," United States, Bureau of Mines, *Report of Investigations*, No. 3221, (Nov. 1933), pp. 1-3.

60 Howard Blaine Burton, "The Romance of Roofing Slate," Part II, *The American Architect*, Vol. 114, No. 2230 (18 Sept. 1918), pp. 333-37; United States Geological Survey. *Mineral Resources of the United States* (hereafter cited as *MRUS*) 1917, Part 2, Non-metals (Washington: Government Printing Office, 1920), pp. 123, 127; ibid., 1929, p. 166.

61 Quebec, Department of Colonization, Mines and Fisheries, *Report on Mining Operations in the Province of Quebec, 1921* (Quebec: King's Printer, 1922), p. 69; ibid., 1922, pp. 92-93.

62 John H. Lutman, *The Historic Heart of London* (London: Corporation of the City of London, 1977), pp. 9-12.

63 Christina Cameron and Janet Wright, op. cit., pp. 13-14, 17.

64 An estimate for covering the new jail at Quebec, 1863, indicated tin would cost $850 more than slate. NA, RG11, Vol. 274, No. 64910, Note of S. Keefer, 27 June 1863. In the early 19th-century Quebec tin was cheaper than slate according to Jean-Pierre Hardy, *Le forgeron et le ferblantier* (Montreal: Les Éditions du Boréal Express, 1978), p. 113. Its popularity continued after it was less economical.

II. Stylistic Trends in the Use of Slate Roofing in Canada

1 Peter Nicholson, *Encyclopedia of Architecture: A Dictionary of the Science and Practice of Architecture, Building, Carpentry, etc.*, Vol. 1 (New York: Johnson, Fry & Co., 1858), p. 431.

2 Edward H. Knight, *Knight's American Mechanical Dictionary*, Vol. 3 (Boston: Houghton, Osgood and Company, 1879), p. 2200; *Canadian Contractors' Hand-Book and Estimator*, third edition (Toronto: Canadian Architect and Builder Press, 1901), p. 120.

3 Peter Nicholson, op. cit., p. 431; J.F. Blondel, *Cours d'architecture ou traité de la décoration, distribution et construction des bâtiments*, Vol. 6 (Paris: Chez la Veuve Desaint, 1777), p. 323.

4 Peter Nicholson, op. cit., pp. 431-33; T. Mellard Reade, "Slate, Slates and Slating," *The Building News*, (16 Jan. 1891), in NA, MG30, B.86, Barnett Engineering Collection, new Vol. 12, old envelope No. 381.

5 Oliver Bowles, "The Technology of Slate," pp. 8-9; J.A. Dresser, "Serpentine Belt of Southern Quebec" in Canada, Department of Mines,

Summary Report of the Geological Survey Branch, 1910 (Ottawa: Queen's Printer, 1911), p. 218.

6   Howard Blaine Burton, "The Romance of Roofing Slate," Part II, *The American Architect*, Vol. 114, No. 2230 (18 Sept. 1918), pp. 334-37.

7   Oliver Bowles, "The Technology of Slate," pp. 3-6.

8   E. Viollet-le-Duc, *Dictionnaire raisonné de l'architecture française du XIe au XVIe siècle*, Vol. 1. (Paris: V.A. Morel & Cie, 1875), pp. 454-55.

9   Peter Nicholson, op. cit., p. 433; Joseph Gwilt, *An Encyclopedia of Architecture, Historical, Theoretical and Practical* (New York: Longman's, Green and Co., 1903), p. 677.

10   E. Viollet-le-Duc, op. cit., pp. 453-54.

11   "Slate and Shingle Patterns," *Carpentry and Building* (Dec. 1884), p. 233.

12   "Scales and Imbrications," *The American Architect and Building News*, Vol. 36, No. 853 (30 April 1892), p. 65.

13   "Slate and Shingle Patterns," *Carpentry and Building* (Dec. 1884), p. 233.

14   Mathilde Brosseau, op. cit., pp. 13-17.

15   A.J. Downing, *Cottage Residences* ... (New York: Wiley and Putman, 1842), p. 107.

16   Ibid., pp. 50, 107.

17   A.J. Downing, *The Architecture of Country Houses* (New York: D. Appleton & Co., 1850), p. 182.

18   John Ruskin, *The Stones of Venice* (New York: John Wiley, 1851), Vol. 1, p. 351.

19   Ibid.

20   Mathilde Brosseau, op. cit., see diamond or lozenge shape shingles, The Grove, Picton, Ontario, p. 102.

21   Henry-Russell Hitchcock, *Architecture, Nineteenth and Twentieth Centuries* (Baltimore: Penguin Books, reprint, 1975), pp. 247-50.

22   Georg Germann, *Gothic Revival in Europe and Britain: Sources, Influences and Ideas* (London: Lund Humphries Publishers, 1972), pp. 118-21.

23   Henry-Russell Hitchcock, op. cit., pp. 276-7; Peter Howell, *Victorian Churches* (London: Royal Institute of British Architects, 1968), pp. 13-19, 23-25.

24   Mathilde Brosseau, op. cit., p. 7; see also Claude T. Bissell, ed., *University College: A Portrait, 1853-1953* (Toronto: University of Toronto Press, 1953), pp. 22-30.

25   NA, RG11, Vol. 841, exhibits 62-63, p. 7, "Semper Paratus" code name for Fuller and Jones, Competition plans of 1859. These original plans called for a metal roof which was subsequently changed to slate.

26 Calvert Vaux, *Villas and Cottages* ... (New York: Harper & Brothers, 1857), p. 61.

27 Samuel Sloan, *Sloan's Homestead Architecture* (Philadelphia: J.B. Lippincott, 1861), pp. 48-49; Henry Hudson Holly, *Holly's Country Seats* (New York: D. Appleton & Co., 1863).

28 The various ecclesiastical emblems are illustrated and named in Henry Hudson Holly, *Church Architecture* (Hartford, Conn.: M.H. Mallory and Company, 1871), pp. 172-73.

29 Mathilde Brosseau, op. cit., pp. 24-25.

30 Luc Noppen, Claude Paulette, and Michel Tremblay, *Québec: trois siècles d'architecture* (Québec: Libre Expression, 1979), pp. 172-73; *Canadian Illustrated News* (hereafter cited as *CIN*), Vol. 4 (29 July 1871), pp. 65-66.

31 "St. Lukes' Church, Waterloo," *CIN*, Vol. 3 (17 June 1871), p. 371; Greenhill, Ralph, Ken MacPherson, and Douglas Richardson, *Ontario Towns* (Ottawa: Oberon Press, 1974), n.p., fig. 9.

32 Begun in 1897 and completed in 1907, the stone cathedral of St. Dunston's with its polychrome slate roof was burned 1913, Irene Rogers, *Charlottetown: The Life in Its Buildings* (Charlottetown: The Prince Edward Island Museum and Heritage Foundation, 1983), p. 116.

33 Christina Cameron and Janet Wright, op. cit., pp. 9-10.

34 Ibid., p. 50.

35 Ibid., p. 52.

36 Isaac H. Hobbs, and son, *Hobb's Architecture: containing designs and ground plans for villas, cottages and other edifices both suburban and rural, adapted to the United States, with rules for criticism and introduction* ... (Philadelphia: J.B. Lippincott & Co., 1875), examples on pp. 36, 38, 46.

37 Marcus Fayette Cummings and Charles Crosby Miller, *Architecture. Designs for street fronts, suburban houses, and cottages, including details for both exterior and interior ... comprising in all 382 designs and 714 illustrations* ... (Troy, New York: Young and Benson, 1865), n.p.

38 Amos Jackson Bicknell, *Supplement to Bicknell's Village Builder containing Eighteen Modern Designs for Country and Suburban Houses of Moderate Cost with Elevations, Plans, Sections* ... (New York: A.J. Bicknell & Co. 1871), Plates 6, 9 and 15.

39 *Canadian Illustrated News*, Vol. 3 (Feb. 1871), p. 103.

40 Archives, Les Soeurs de la Congrégation de Notre-Dame. Agreement between the Sisters of the Congregation of Notre Dame and Mr. G.W. Reed for Roofing Maison-Mère Villa-Maria, Montreal, 15 July 1878.

41 See Christina Cameron and Janet Wright, op. cit., pp. 92-95.

42 "Slate and Shingle Patterns," *Carpentry and Building* (December, 1884), p. 233; "Roofing Materials," *The American Architect*, Vol. 129, No. 2488 (5 Jan. 1926), pp. 111-12.

43 Mark Girouard, *Sweetness and Light: The 'Queen Anne' Movement 1860-1900* (Oxford: Clarendon Press, 1977), pp. 12-13; Walter C. Kidney, *The Architecture of Choice: Eclecticism in America 1880-1930* (New York: George Braziller, 1974), pp. 4-9; for a detailed discussion of the "Queen Anne" in this country, see Leslie Maitland, "The Queen Anne Revival Style in Canadian Architecture," Manuscript in progress (Ottawa: Environment Canada, Parks Service, 1986).

44 J.J. Stevenson, *House Architecture*, Vol. 2 (London: Macmillan and Co., 1880), p. 206.

45 William Morris, *The Collected Works of William Morris*, Vol. 22, (London: Longman's, Green and Company, 1914), p. 407.

46 H. Hudson Holly, *Modern Dwellings in Town and Country Adapted to American Wants and Climate* (New York: Harper & Brothers, Publishers, 1878), p. 24.

47 William T. Comstock, *American Cottages* (New York: William T. Comstock, 1883), Plates XVI, XIX.

48 "Slate and Shingle Patterns," *Carpentry and Building* (December 1884), p. 233.

49 "Our Illustrations," *CA&B*, Vol. 4, No. 1 (Jan. 1891), p. 4.

50 Henry-Russell Hitchcock, op. cit., pp. 312-26.

51 J.J. Stevenson, op. cit., p. 211.

52 H. Hudson Holly, *Modern Dwellings* p. 64.

53 "Redpath Library, McGill University," *CA&B*, Vol. 8, No. 8 (August 1895), pp. 96-97; "Legislative Buildings, Victoria, B.C.," ibid., Vol. 11, No. 7 (July 1898), p. 120; Eric Arthur, *From Front Street to Queen's Park: The Story of Ontario's Parliament Buildings* (Toronto: McClelland and Stewart, 1979), pp. 77, 81; "Public Library, London, Ontario," *CA&B*, Vol. 7, No. 12 (December 1894), p. 156.

54 Harold D. Kalman, *The Railway Hotels and the Development of the Château Style in Canada* (Victoria: University of Victoria Maltwood Museum, 1968), pp. 15-16 and figure pages unnumbered.

55 Martin Segger and Douglas Franklin, *Victoria: A Primer of Regional History in Architecture, 1843-1929* (Watkins Glen, N.Y.: A Pilgrim Guide to Historic Architecture, 1979), p. 285.

56 Henry-Russell Hitchcock, op. cit., pp. 318-26.

57 Ibid., pp. 369-79.

58 Walter C. Kidney, op. cit., p. 32.

59 "Slate," *MRUS*, 1906 (Washington: Government Printing Office, 1907), pp. 1001, 1004.

60 "Residential Design in Canada," *Construction*, Vol. 3, No. 8 (July 1910), pp. 76-80.

61 Philip J. Turner, "Houses at Montreal, Quebec," *Construction*, Vol. 8, No. 6 (June 1915), pp. 265-73; "Recent Houses in Montreal and West-mount," ibid., Vol. 9, No. 6 (June 1916), pp. 198-200.

62 "Attractiveness in Slate Roofs," *The Contract Record*, Vol. 22, No. 46 (November 1908), p. 24.

63 Ibid.; National Slate Association, *Slate Roofs* (Reprint of 1926 edition by Vermont Structural Slate Co. Inc., Fair Haven, Vermont, 1977), [here-after cited as National Slate Association, *Slate Roofs*], pp. 5-6, 58-59.

64 "Attractiveness in Slate Roofs," *The Contract Record*, p. 24.

65 M.H. Baillie Scott, "The Art of Building," *Construction*, Vol. 3, No. 4 (April 1910), p. 97.

66 Ibid.

67 "Mountain Side House at Westmount, Montreal," *Construction*, Vol. 10, No. 4 (April 1917), pp. 162-63.

68 "Examples of Recent Domestic Work," *Construction*, Vol. 12, No. 6 (June 1919), pp. 177-78.

69 NA, MG31, B.38, H.L. Fetherstonhaugh Collection, Project No. 231, Building for A.W. Patterson, Drummond St., Montreal, June 1929.

70 Howard Blaine Burton, "The Romance of Roofing Slate, Part II, "*The American Architect*, Vol. 114, No. 2230 (Sept. 1918), pp. 333-37.

71 Oliver Bowles, "Recent Progress in Slate Technology," United States, Bureau of Mines, Reports of Investigations, Serial No. 2766 (August 1926), p. 6.

72 NA, MG31, B38, H.L. Fetherstonhaugh Collection, Project No. 228, Residence for Gregor Barclay, Redpath Crescent, Montreal, December 1927.

73 "An Unusual Small Stone House," *Canadian Homes and Gardens*, Vol. 15, No. 8 (August 1927), p. 24; "Montreal Badminton and Squash Club," *Construction*, Vol. 21, No. 1 (Jan. 1928), pp. 5-11.

III. Canadian Practice in Laying Slate Roofs

1 "Who Does the Slate Roofing in Your Town?" *Metal Worker, Plumber and Steam Fitter*, Vol. 90 (April 1918), pp. 522-23.

2 Compare specifications for the East Block of Parliament Hill, 1863, the Ottawa Post Office, 1871, Walmer Road Baptist Church, Toronto, 1892 and W.R.G. Holt Residence, Montreal, 1926 in J. Daniel Livermore, *Departmental Buildings, Eastern Block, Parliament Hill, Ottawa. An*

Historical Report on the East Block Restoration (Ottawa: Department of Public Works, 1973), App. D; NA, RG11, Vol. 3920, p. 227; "Walmer Road Baptist Church, Toronto," *CA&B*, Vol. 5, No. 5 (June 1892), p. 62; NA, MG31, B.38, H.L. Fetherstonhaugh Collection, Vol. 1, Box 1, Project 229, Specifications, p. 33.

3   "Who Does the Slate Roofing in Your Town?" op. cit., pp. 522-23.

4   Philip C. Marshall, op. cit., p. 77.

5   Peter Nicholson, op. cit., pp. 431-43; Joseph Gwilt, op. cit., pp. 520, 676-80; J.C. Loudon, *An Encyclopedia of Cottage, Farm and Villa Architecture and Furniture* (London: Longman, Brown, Green and Longman's, 1846), pp. 119, 564, 1150-51, 1307.

6   "Rule for Measurements," *CA&B*, Vol. 10, No. 9 (Sept. 1897), p. 181; "Roofing," *The American Architect and Building News*, Vol. 36, No. 860 (18 June 1892), p. 175.

7   For example, A.J. Downing, *Hints to Persons About Building in the Country* (New York: John Wiley, 1859); Samuel Sloan, op. cit., pp. 48-49, 51; John Bullock, *The Rudiments of the Art of Building* (New York: Stringer and Townsend, 1853), pp. 133-35.

8   "Lead, Copper and Zinc for Roofing," *CA&B*, p. 205. Architect Julian Smith, Restoration Services Division, Parks Service, Environment Canada, provided the interesting fact that Hodgson was Canadian, from Collingwood, Ontario.

9   "Slates and Slating," *CA&B*, Vol. 9, No. 6 (June 1896), p. 88; "Specifications for Roofing Slate," *The Canadian Engineer*, Vol. 59, No. 2 (8 July 1930), p. 136.

10 The two examples of slate industry literature that the author was able to obtain were both productions of a Bangor, Pennsylvania slate company: The first located in the University of Vermont, Wilbur Collection, was John Galt & Sons, Bangor Excelsior Slate Company, *Roofing Slate 1896: What to Use, How to Use it* (hereafter cited as *Roofing Slate 1896*). The second booklet found in the Barnett Collection of the National Archives was also by the Bangor Excelsior Slate Company and entitled *Slate Roofs* (third edition, 1897).

11 National Slate Association, *Slate Roofs*.

12 Oliver Bowles, "Recent Progress in Slate Technology," op. cit., pp. 8-9; *Canadian Mining Journal*, Vol. 47, No. 43 (Oct. 1926), pp. 1017-18, 1024.

13 "Slates and Slating," *CA&B*, Vol. 9, No. 6 (June 1896), p. 88; "Slates," ibid., Vol. 11, No. 5 (May 1898), p. 92.

14 NA, RG11, Vol. 274, No. 64910, P. Gauvreau to T. Trudeau, 23 June 1863.

15 Peter Nicholson, op. cit., pp. 432-33; J.C. Loudon, op. cit., pp. 1307, 1150-51.

16 Canada, Department of Consumer and Corporate Affairs, Patent Office, Joseph Scobell, Application for Letters Patent of Invention for a new and improved method of covering roofs with slates, 8 June 1854; *Patents of Canada from 1849 to 1855*, Vol. 11 (Toronto: Lovell & Gibson, 1865), No. 463.

17 Joseph Gwilt, op. cit., p. 677.

18 See ad for Williams flat slate roof, William & Co., Toronto in *CA&B*, Vol. 15, No. 12 (Dec. 1902), p. xvi.

19 "The National Research Building, Ottawa," *Construction*, Vol. 25, No. 8 (August 1932), p. 178; Canada, Department of Public Works, "Specification for National Research Laboratories, Ottawa, Ontario, Sproatt and Rolph, Architects," July 1929.

20 NA, RG11, Vol. 274, No. 64910, P. Gauvreau to T. Trudeau, 23 June 1863; "Roof Construction"; *CA&B*, Vol. 1, No. 7 (July 1888), p. 6;

21 "Roof Construction," *CA&B*, p. 6.

22 Peter Nicholson, op. cit., pp. 431-33; Joseph Gwilt, op. cit., pp. 676-79; "Roofing," *The American Architect and Building News*, pp. 175-76.

23 NA, MG30, B.86, Barnett Engineering Collection, Vol. 12, Catalogue 2, Bangor Excelsior Slate Co., *Slate Roofs*, third edition (1897), p. 11.

24 Genuine Bangor Slate Co., "Slate and Its Uses," (Easton, Penn.: Genuine Bangor Slate Co., 1907), pp. 24, 32; National Slate Association, *Slate Roofs*, p. 34.

25 "These heavy slates are said to be used principally for roofing on huge steel buildings of the class now being built in the larger cities." Canada, Geological Survey, *Annual Report*, 1910, Sessional Paper No. 26 (Ottawa: Queen's Printer, 1911), p. 218; "Neither large sizes nor thicknesses can be successfully carried on any but heavy supporting walls." Howard Blaine Burton, op. cit., p. 337.

26 John Bullock, op. cit., p. 133; Joseph Gwilt, op. cit., p. 676.

27 Peter Nicholson, op. cit., p. 432.

28 Bangor Excelsior Slate Company, *Roofing Slate 1896*, p. 5. National Slate Association, *Slate Roofs*, p. 40.

29 Documents Relating to the Construction of the Parliamentary and Departmental Buildings at Ottawa (Quebec: Department of Public Works, 1862), pp. 70-71.

30 NA, WO55, Vol. 887, pp. 576-79; "Specifications for Walmur Road Baptist Church, Toronto," *CA&B*, p. 61.

31 National Slate Association, *Slate Roofs*, p. 40.

32 Diagonal sheathing was specified for slating at Fort Needham, Halifax, NA, WO55, Vol. 887, pp. 576-79; Double boarding was noted in the 1920s residences of the H.L. Fetherstonhaugh Collection, NA, MG31, B.38, Vol. 1, Projects 228, 229, 233.

33 Claims such as Gwilt's that felt was impervious to rain, snow or frost were denied by the National Slate Association who saw its chief purpose as a temporary covering and slight insulator. Joseph Gwilt, op. cit., p. 676; National Slate Association, *Slate Roofs*, pp. 23-24.

34 See, for example, H.L. Fetherstonhaugh Collection, NA, MG31, B.38, Vol. 1, Project 228.

35 Joseph Gwilt, op. cit., p. 676; specifications never mentioned battens in conjunction with board undersheathing and felt.

36 "The National Research Building, Ottawa," *Construction*, p. 178; "McMaster University, Hamilton, Ontario," *Construction*, Vol. 23, No. 11 (Nov. 1930), pp. 375, 377.

37 See Peter Nicholson, op. cit., p. 432.

38 James H. Bowen, *Report upon Buildings, Building Materials and Methods of Building, Paris Universal Exposition, 1867* (Washington: Government Printing Office, 1869), p. 52; "Roofing," *The American Architect and Building News*, p. 175.

39 "Slates," *CA&B*, Vol. 11, No. 5 (May 1898), p. 92.

40 For further detail on flashings see National Slate Association, *Slate Roofs*, pp. 27-33.

41 Peter Nicholson, op. cit., p. 432.

42 Bangor Excelsior Slate Co., *Slate Roofs*, third edition, 1897, p. 11; National Slate Association, *Slate Roofs*, p. 23.

43 "Slates and Slating," *CA&B*, Vol. 9, No. 6 (June 1896), p. 88.

44 NA, RG11, Vol. 3920, p. 227, Specifications for the Ottawa Post Office, 1871.

45 Mary K. Cullen, op. cit., Appendix B, Specifications annexed to the builders contracts, 1842, p. 122.

46 "Slates," *CA&B*, p. 92.

47 NA, MG31, B.38, Vol. 1, Box 1, Project 229.

48 NA, RG11, Vol. 3921, p. 74.

49 Joseph Gwilt, op. cit., p. 676; according to C.F. Innocent pointing mortar had to be haired or it would not stick to the laths, "Points on the Supervision of Buildings," *The Contract Record*, Vol. 22, No. 34 (19 August 1908), n.p.

50 Mary K. Cullen, op. cit., p. 122.

51 "On Slate Roofs," *CA&B*, Vol. 12, No. 6 (June 1899), p. 123.

52 *Canadian Contractor's Hand-Book and Estimator*, third edition (Toronto: Canadian Architect and Builder Press, 1901), p. 121.

53 Joseph Gwilt, op. cit., p. 679; "A very beautiful and ornamental ridge covering is now manufactured of slate ...," Samuel Sloan, *Sloan's Homestead Architecture* p. 51.

54 "Example of Recent Domestic Work," *Construction*, Vol. 12, No. 6 (June 1919), p. 178.

55 The purpose of making valleys wider at the bottom "to allow for snow sliding" was expressed in NA, RG11, Vol. 3909, Specifications for Extension of Western Block of Departmental Buildings, Ottawa, Ontario, 1875.

56 National Slate Association, *Slate Roofs*, pp. 22-23.

IV. Sources of Roofing Slate, Past and Present

1 W.A. Parks, *Report on the Building and Ornamental Stones of Canada*, Vol. 3, Canada, Department of Mines, Mines Branch, Publication No. 279 (Ottawa: King's Printer, 1914), pp. 235; John A. Dresser, "On the Slate Industry in Southern Quebec," *The Canadian Mining Journal*, Vol. 32 (15 Sept. 1911), pp. 584-85.

2 John A. Dresser, op. cit., pp. 585-86; Canada, Geological Survey, *Annual Report*, 1888-89, Part K, R.W. Ells, "On the Mineral Resources of the Province of Quebec," pp. 128K-131K.

3 *Descriptive Catalogue of a Collection of the Economic Minerals of Canada and Notes on a Stratigraphical Collection of Rocks. Philadelphia International Exhibition, 1876* (Montreal: Lovell Printing and Publishing, 1876), p. 107.

4 Ibid., p. 107; Canada, Geological Survey, *Annual Report*, 1888-89, Part K, p. 131K.

5 "The Duty on Building Materials," *CA&B*, Vol. 10, No. 3 (March 1897), p. 44.

6 NA, RG89, Vol. 34, Geological and Natural History Survey, Mining Returns, 1892, 1893, 1895, 1897.

7 "The Stone Resources of Newfoundland," *CA&B*, Vol. 15, No. 11 (Nov. 1902), pp. 170-71; G.F. Carr, op. cit., pp. 134-37.

8 Canada, Geological Survey, *Report of Progress for the Year 1852-3* (Quebec: John Lovell, 1854), pp. 72-73; Robert Bell, op. cit., p. 6.

9 *Descriptive Catalogue of a Collection of the Economic Minerals of Canada and Notes on a Stratigraphical Collection of Rocks. Philadelphia International Exhibition, 1876*, op. cit., p. 108; Canada, Geological Survey, *Annual Report*, 1888-89, Part K, p. 129K.

10 John A. Dresser, op. cit., p. 587.

11 Canada, Geological Survey, *Annual Report*, 1888-89, Part K, p. 131.

12 NA, RG87, Vol. 34, Records of the Mineral Resources Branch, Slate Production Information, 1885-1919.

13 "Monopoly of Slate Company," *CA&B*, Vol. 11, No. 6 (June 1889), p. 62.

14 W.A. Parks, op. cit., p. 240.

15 NA, RG16, A-3, Vol. 808, Customs Tariff, being a consolidation of acts 1879 to 1886, item No. 129; 49 Vic. Cap. 33, An Act respecting Customs Duties, 1886, Item No. 380; Analytical Index to Customs Tariffs of the Dominion of Canada as in force July 1893.

16 "The Duty on Building Materials," *CA&B*, Vol. 10, No. 3 (March 1897), p. 44.

17 NA, RG16, A-3, Vol. 809, The Customs Tariff 1897, Item No. 197.

18 "Roof coverings," *CA&B*, Vol. 13, No. 10 (Oct. 1900), p. 186.

19 Statistics on the annual quantity and value of roofing slate production in the United States 1879-1912 appear in T. Nelson Dale, "Slate in the United States," p. 200. Canadian statistics can be consulted in Chapter I, Tables 3 and 4.

20 Ibid.

21 United States, Geological Survey, *MRUS*, 1906, Part 11, p. 1003.

22 *Tables of the Trade and Navigation of the Dominion of Canada*, Table No. 1: General Statement (by Countries and Provinces) of the Total Quantities and Values of Merchandise Imported ... Roofing Slate, 30 June 1882, p. 313; ibid., 1883, p. 318; ibid., 1884, p. 321; ibid., 1885, p. 324; ibid., 1886, p. 344; ibid., 1887, p. 326; ibid., 1888, pp. 341-42; ibid., 1889, p. 329; ibid., 1890, p. 343; ibid., 1891, p. 194; ibid., 1892, p. 198; ibid., 1893, p. 198; ibid., 1894, p. 205; ibid., 1895, p. 262; ibid., 1896, p. 205; ibid., 1898, p. 134.

23 "Building Materials," *CA&B*, Vol. 1, No. 7 (July 1888), ibid., (Sept. 1888), ibid., (March 1889).

24 T. Nelson Dale, *Slate in the United States*, pp. 193-201.

25 United States, Geological Survey, *MRUS*, 1912, Part II, Nonmetals, T. Nelson Dale, "The Commercial Qualities of the Slates of the United States and their Localities," (hereafter cited as T. Nelson Dale, "The Commercial Qualities of the Slates of the United States"), pp. 698-700.

26 Mansfield Merriman, "The Slate Regions of Pennsylvania," *Stone*, Vol. 17, No. 2 (July 1898), p. 80.

27 T. Nelson Dale, "Slate in the United States," pp. 105-8.

28 Mansfield Merriman, op. cit., p. 83; T. Nelson Dale, "The Commercial Qualities of the Slates of the United States," p. 699.

29 T. Nelson Dale, "Slate in the United States," pp. 74-83; ibid., "The Commercial Qualities of the Slates of the United States," pp. 695-96.
30 Mansfield Merriman, op. cit., p. 77.
31 T. Nelson Dale, "The Commercial Qualities of the Slates of the United States," pp. 698-700, 704-6; Mansfield Merriman, op. cit., p. 81.
32 "Building Materials," *CA&B*, p. 11; ibid., Vol. 2, No. 8 (August 1889), p. 111; ibid., Vol. 3, No. 3 (March 1890), p. 10.
33 Philip C. Marshall, op. cit., pp. 79 and 85; Mansfield Merriman, op. cit., p. 79.
34 T. Nelson Dale, "The Commercial Qualities of the Slates of the United States," pp. 705-6.
35 "Slate in Canada," *CA&B*, Vol. 12, No. 6 (June 1899), p. 127.
36 T. Nelson Dale, "The Commercial Qualities of the Slates of the United States," pp. 702-4. Milford D. Whedon, "The New York-Vermont Slate Belt," *Stone*, Vol. 28 (1909), pp. 214-18.
37 Compare outputs of roofing slate for the states of Pennsylvania and Vermont 1901-13 in T. Nelson Dale, "Slate in the United States," pp. 197-200.
38 United States, Geological Survey, *MRUS*, Part II, 1915, p. 23; Milford D. Whedon, op. cit., p. 217.
39 United States, Geological Survey, *MRUS*, Part II, 1915, p. 23.
40 Ibid., 1908, p. 526; 1909, p. 565; 1910, p. 638; 1911, p. 734; 1912, p. 687; 1915, p. 23.
41 Ibid., 1917, p. 127.
42 "Roof coverings," *CA&B*, p. 186.
43 T. Nelson Dale, "Slate in the United States," pp. 195-99.
44 United States, Geological Survey, *MRUS*, 1904, p. 830, Exports of slate from the United States, showing ports and customs districts from which and to which sent, in the fiscal years 1895-1904.
45 Names and addresses of the Canadian and the major U.S. roofing slate quarries are:
    1) Island Tile and Slate Company,
        Nutcove, Trinity Bay,
        Newfoundland
        (709) 663-5566.
    2) Buckingham-Virginia Slate Corp.,
        Box 11002, 4110 Fitzhugh Ave.,
        Richmond, VA. 23230
        (804) 355-4351.
    3) Evergreen Slate Co.,
        68 East Potter Ave.,

Granville, N.Y. 12832
(518) 642-2530.

4) Hilltop Slate Co.,
   Rt. 22A,
   Middle Granville
   N.Y. 12849
   (518) 642-2270.

5) Rising & Nelson Slate Co.,
   West Pawlett, VT. 05775
   (802) 645-0150.

6) Structural Slate Co.,
   222 E. Main Street,
   Pen Argyl, PA. 18072
   (215) 863-4141.

7) Vermont Structural Slate Co.,
   P.O. Box 98,
   Fairhaven, VT. 05743
   (800) 343-1900 or in VT. (802) 265-4933.

46 *Annual Book of ASTM Standards, 1983.* Vol. 04.08, Designation C406-58 (Re-approved 1981), pp. 25-27.

47 Ibid., Methods C120, C121, and C217, pp. 11-13, 15, 19-21.

48 "Slate Roofs," *Old House Journal*, Vol. 8, No. 5 (May 1980), p. 49.

49 David Hein, "Roofing with Slate," *Fine Homebuilding*, No. 20 (April/May 1984), p. 239.

Selected Bibliography

Adams, Blaine
"The Construction and Occupation of the Barracks of the King's Bastion at Louisbourg." *Canadian Historic Sites: Occasional Papers in Archaeology and History*, No. 18 (1978), pp. 59-147. Ottawa.

American Architect and Building News, The (Boston)
"Scales and Imbrications." Vol. 36, No. 853 (30 April 1892), pp. 65-67.
———. "Roofing." Vol. 36, No. 860 (18 June 1892), pp. 175-76.
———. "Roofing Materials." Vol. 129, No. 2488 (5 January 1926), pp. 111-12.

Annual Book of ASTM Standards, 1983
Section 4 — Construction. Vol. 04.08 — Soil and Rock, Building Stones. ASTM, Philadelphia, 1983.

Arthur, Eric
From Front Street to Queen's Park: The Story of Ontario's Parliament Buildings. McClelland and Stewart, Toronto, 1979.

Baillairgé, G.F.
"Description and Cost of the Public Buildings Constructed or Improved by the Department of Public Works." Canada, Department of Public Works, *Annual Report*, 1867, App. No. 23, pp. 248-84. Ottawa.

Bangor Excelsior Slate Company
Roofing Slate 1896: What Kind to Use, How to Use it and Showing its Superior Advantages over other Roofing Material. John Galt & Sons, Bangor Excelsior Slate Co., Bangor, PA., 1896.
———. *Slate Roofs.* Third edition, 1897. Bangor Excelsior Slate Co., Easton, PA., 1897.

Bell, Robert
Roofing Slate as a Source of Wealth to Canada: A Visit to the Walton Slate Quarry. A paper read before the Natural History Society, Melbourne, Canada East, 8 October 1863.

Bicknell, Amos Jackson
Supplement to Bicknell's Village builder, containing Eighteen Modern Designs for Country and Surburban Houses of Moderate Cost with Elevations, Plans, Sections, and a Variety of Details ... A.J. Bicknell and Co., New York, 1871.

Bissell, Claude T., ed.
University College: A Portrait 1853-1953. University of Toronto Press, Toronto, 1953.

Blondel, J.F.
Cours d'architecture ou traité de la décoration, distribution et construction des bâtiments. Chez la Veuve Desaint, Paris, 1777, 6 vols. Vol. 6.

Bowles, Oliver
"The Technology of Slate." United States, Department of the Interior, Bureau of Mines, *Bulletin No. 218.* Government Printing Office, Washington, 1922.
———. "The Characteristics of Slate." *American Society for Testing Materials, Proceedings of the 26th Annual Meeting,* Part 2, Technical Papers. (June 1923), pp. 524-72.
———. "Recent Progress in Slate Technology." United States, Department of the Interior, Bureau of Mines, *Reports of Investigations,* No. 2766. (August 1926).
———. "The Wire Saw in Slate Quarrying," United States, Department of the Interior, Bureau of Mines, *Technical Paper 469.* Government Printing Office, Washington, 1930.
———. "Consumption Trends in the Roofing Slate Industry." United States, Department of Interior, Bureau of Mines, *Reports of Investigations,* No. 3221. (November 1933).

Bowen, James H.
Report upon Buildings, Building Materials and Methods of Building, Paris Universal Exposition, 1867. Government Printing Office, Washington, 1869.

Brosseau, Mathilde
"Gothic Revival in Canadian Architecture." *Canadian Historic Sites: Occasional Papers in Archaeology and History,* No. 25 (1980). Ottawa.

Bullock, John, ed.
The Rudiments of the Art of Building. Stringer and Townsend, New York, 1853.

Burton, Howard Blaine
"The Romance of Roofing Slate." Part II. *The American Architect*, Vol. 114, No. 2230 (18 Sept. 1918), pp. 332-37. Boston.

Cameron, Christina, and Janet Wright
"Second Empire Style in Canadian Architecture." *Canadian Historic Sites: Occasional Papers in Archaeology and History*, No. 24 (1980). Ottawa.

Canada. Dominion Bureau of Statistics.
Canadian Mineral Statistics 1886-1956; Mining Events 1604-1956. Reference Paper No. 168. Queen's Printer, Ottawa, 1957.

Canada. Geological Survey.
"Report of Progress ..." *Journal of the Legislative Assembly of the Province of Canada,* Queen's Printer, Quebec, Toronto and Ottawa, 1843-84. No report, [1858-62].
——. *Report of Progress from the Commencement to 1863.* Dawson Brothers, Montreal, 1865. *Annual Reports (New Series).* Queen's Printer, Ottawa, 1884-1904.
——. *Summary Report.* Queen's Printer, Ottawa, 1905-17.

Canada. National Archives.
MG13, WO55, Vol. 863, R. Bydon to General Mann, 19 June 1826; Vol. 1551(7), North American Provinces Commissioner's Report, 9 Sept. 1825; Vol. 887, Specification for slate roof, Fort Needham, Halifax.
. MG24, D.11, Phoenix Assurance Company of London, Vol. 1, Visits to Canada and Newfoundland 1808-9; Vol. 2, Visits to Canada and Newfoundland, 1845-6.
——. MG30, B.86, Barnett Engineering Collection, new Vol. 12, old envelope No. 381 and catalogue 2.
——. MG31, B.38, H.L. Fetherstonaugh Collection, Project Nos. 228, 229, 231, 233.
——. RG11, Public Works Records, Vol. 372, No. 52201, Walton Slate Quarries; Vol. 274, No. 64910, Note of S. Keefer, 27 June 1863; Vol. 841, exhibits 62-63, Semper Paratus to F.P. Rubidge, proposal for Parliament Buildings, n.d.; Series 14, Vols. 3909-3929, 3931, 3939, 3010, 3030, 3035 and 3041, Specifications for Public Buildings, 1871-1907.
——. RG16, A-3, Customs Tariff, Vol. 808, Items No. 129, 380; Vol. 809, Item No. 197.

——. RG87, Vol. 34, No. 140. Records of the Mineral Resources Branch, Slate Production Information, 1885-1919.

Canadian Architect and Builder (Toronto)
1888-1908
"Roof Construction." Vol. 1, No. 7 (July 1888), p. 6.

——. "Building Materials." Vol. 1, No. 7 (July 1888), p. 11; Vol. 1, No. 9 (Sept. 1888); Vol. 2, No. 3 (March 1889); Vol. 2, No. 8 (August 1889), p. 111; Vol. 3, No. 3 (March 1890), p. 10.

——. "Monopoly of Slate Company." Vol. 2, No. 6 (June 1889), p. 62.

——. "The Canadian Slate Industry." Vol. 2, No. 7 (July 1889), p. 82.

——. "Our Illustrations." Vol. 4, No. 1 (Jan. 1891), p. 4.

——. "Walmer Road Baptist Church, Toronto." Vol. 5, No. 5 (June 1892), p. 61.

——. "Public Library, London, Ontario." Vol. 7, No. 12 (Dec. 1894), p. 156.

——. "Redpath Library, McGill University." Vol. 8, No. 8 (Aug. 1895), pp. 96-97.

——. "Slates and Slating." Vol. 9, No. 6 (June 1896), p. 88.

——. "Lead, Copper and Zinc for Roofing." Vol. 9, No. 12 (Dec. 1896), pp. 202-3.

——. "The Duty on Building Materials." Vol. 10, No. 3 (March 1897), p. 44.

——. "Rule for Measurements." Vol. 10, No. 9 (Sept. 1897), p. 181.

——. "Slates." Vol. 11, No. 5 (May 1898), p. 92.

——. "Legislative Buildings, Victoria, B.C." Vol. 11, No. 7 (July 1898), p. 120.

——. "On Slate Roofs." Vol. 12, No. 6 (June 1899), p. 123.

——. "Slate in Canada." Vol. 12, No. 6 (June 1899), pp. 127-28.

——. "Roof Coverings." Vol. 13, No. 10 (Oct. 1900), p. 210.

——. "Slate Deposits in Newfoundland." Vol. 14, No. 3 (March 1901), p. 63.

——. "The Slate Industry in Newfoundland." Vol. 15, No. 7 (July 1902), p. 109.

——. "The Stone Resources of Newfoundland." Vol. 15, No. 11 (Nov. 1902), pp. 170-71.

——. "Scarcity of Slate." Vol. 16, No. 185 (May 1903), p. 1.

——. "Demand for Roofing Slates." Vol. 16, No. 186, (June 1903), p. 107.

——. "The Slate Industry." Vol. 16, No. 189 (Sept. 1903), p. 144.

——. "Montreal Letter." Vol. 17, No. 196 (April 1904), p. 73.

Canadian Contractor's Handbook and Estimator
Third edition. Canadian Architect and Builder Press, Toronto, 1901.

Canadian Engineer, The (Toronto)
"Specifications for Roofing Slate." Vol. 59, No. 2 (8 July 1930), p. 136.

Canadian Illustrated News (Montreal)
"Government House, Toronto." Vol. 3 (18 Feb. 1871), p. 103.
——. "St. Lukes' Church, Waterloo." Vol. 3 (17 June 1871), pp. 371-72.
——. "Military College, Kingston." Vol. 15 (9 June 1877) and Vol. 18, (14 Sept. 1878).

Carpentry and Building (New York)
"Slate and Shingle Patterns." (Dec. 1884), p. 233.

Carr, G.F.
The Industrial Minerals of Newfoundland. Mines Branch Publication No. 855. Department of Mines and Technical Surveys, Ottawa, 1955.

Comstock, William T.
American Cottages ... William T. Comstock, New York, 1883.

Construction (Toronto)
1908-1934
"Residential Design in Canada." Vol. 3, No. 8 (July 1910), pp. 76-80.
——. "Mountain Side House at Westmount, Montreal." Vol. 10, No. 4 (April 1917), pp. 162-63.
——. "Examples of Recent Domestic Work." Vol. 12, No. 6 (June 1919), pp. 177-78.
——. "Colour Effects in Slate." Vol. 16, No. 7 (July 1923), p. 262.
——. "Montreal Badminton and Squash Club." Vol. 21, No. 1 (Jan. 1928), pp. 5-11.
——. "McMaster University, Hamilton, Ontario." Vol. 23, No. 11 (Nov. 1930), pp. 375-77.
——. "The National Research Building, Ottawa." Vol. 25, No. 8 (Aug. 1932), p. 178.

Contract Record, (The)
"Points on the Supervision of Buildings." Vol. 22, No. 34 (19 Aug. 1908), n.p.
——. "Attractiveness in Slate Roofs." Vol. 22, No. 46 (Nov. 1908), p. 24.
——. "Prepared Roofing." Vol. 25, No. 31 (Aug. 1911), pp. 48-49.

Cullen, Mary K.
"A History of the Structure and Use of Province House, Prince Edward Island, 1837-1977." Manuscript Report Series No. 211. Parks Canada, Ottawa, 1977.

Cummings, Marcus Fayette, and Charles Crosby Miller
Architecture. Designs for street fronts, suburban houses, and cottages, including details for both exterior and interior ... comprising in all 382 designs and 714 illustrations ... Young and Benson, Troy, New York, 1865.

Dale, T. Nelson
"The Commercial Qualities of the Slates of the United States and their Localities." United States. Geological Survey. *Mineral Resources of the United States, 1912*, Part 2, pp. 693-707.

Dale, T. Nelson et al.
"Slate in the United States." United States. Department of the Interior. *Geological Survey Bulletin No. 586.* Government Printing Office, Washington, 1914.

Descriptive Catalogue of a Collection of the Economic Minerals of Canada and Notes on a Stratigraphical Collection of Rocks. Philadelphia International Exhibition, 1876.
Lovell Printing and Publishing, Montreal, 1876.

Downing, Andrew Jackson
Cottage Residences ... Wiley and Putman, New York, 1842.
——. *The Architecture of Country Houses* ... D. Appleton & Co., New York, 1850.
——. *Hints to Persons About Building in the Country.* John Wiley, New York, 1859.

Dresser, J.A.
"Serpentine Belt of Southern Quebec," in Canada, Department of Mines, *Summary Report of the Geological Survey Branch, 1910.* Queen's Printer, Ottawa, 1911, pp. 217-18.
——. "On the Slate Industry in Southern Quebec." *Canadian Mining Journal*, Vol. 32 (15 Sept. 1911), pp. 584-90.

Fauteaux, Joseph-Noel
Essai sur l'industrie au Canada sous le régime français. Ls-A. Proulx, Quebec, 1927, 2 vols. Vol. 1.

Genuine Bangor Slate Co.
Slate and Its Uses. Second edition, Genuine Bangor Slate Co., Easton, Pennsylvania, 1907. (Booklet in the possession of the Historical Society of Pennsylvania.)

Germann, Georg
Gothic Revival in Europe and Britain: Sources, Influences and Ideas. Lund Humphries Publishers, London, 1972.

Girouard, Mark
Sweetness and Light: The 'Queen Anne' Movement 1860-1900. Clarendon Press, Oxford, 1977.

Greenough, John Joseph
"The Halifax Citadel 1825-60: A Narrative and Structural History." *Canadian Historic Sites: Occasional Papers in Archaeology and History*, No. 17 (1977). Ottawa.
———. "South Magazine, Halifax Citadel, A Structural History." Manuscript Report Series, No. 223. Parks Canada, Ottawa, 1977.

Gwilt, Joseph
An Encyclopedia of Architecture, Historical, Theoretical and Practical. Longman's, Green and Co., New York, 1903 edition.

Hein, David
"Roofing with Slate." *Fine Homebuilding*, No. 20 (April/May 1984), pp. 38-43.

Hitchcock, Henry-Russell
Architecture, Nineteenth and Twentieth Centuries. Penguin Books, Baltimore, 1975 reprint.

Hobbs, Isaac H., and son
Hobbs' Architecture: containing designs and ground plans for villas, cottages and other edifices both suburban and rural, adapted to the United States, with rules for criticism and introduction ... J.B. Lippincott & Co., Philadelphia, 1875.

Holly, Henry Hudson
Holly's Country Seats. D. Appleton & Co., New York, 1863.
———. *Church Architecture.* M.H. Mallory and Company, Hartford, Connecticut, 1871.
———. *Modern Dwellings in Town and Country Adapted to American Wants and Climate.* Harper & Brothers, New York, 1878.

Howell, Peter
Victorian Churches. Royal Institute of British Architects, London, 1968.

Kalm, Peter
Travels in North America. Dover Publications, New York, 1966. Vol. 1.

Kalman, Harold D.
The Railway Hotels and the Development of the Château Style in Canada.
University of Victoria Maltwood Museum, Victoria, B.C., 1968.

Kidney, Walter C.
The Architecture of Choice: Eclecticism in America 1800-1930. George Braziller, New York, 1974.

Knight, Edward H.
Knight's American Mechanical Dictionary. Houghton, Osgood and Company, Boston, 1879. Vol. 3.

Laframboise, Yves
L'architecture traditionnelle au Québec. Les Éditions de l'Homme Ltée., Montreal, 1975.

Lambert, John
Travels through Canada and the United States of North America in the years 1806, 1807 and 1809. C. Craddock and W. Jay, London, 1814. 2 vols. Vol. 1.

Les Soeurs de la Congrégaton de Notre-Dame, Archives (Montreal)
"Agreement between the Ladies of the Congregation of Notre Dame and Mr. G.W. Reed for roofing Maison-mère Villa-Maria," Montreal, 15 July 1878.

Lindsay, Jean
A History of the North Wales Slate Industry. David and Charles, London, 1974.

Livermore, J. Daniel
Departmental Buildings, Eastern Block, Parliament Hill, Ottawa. An Historical Report on the East Block Restoration. Canada, Department of Public Works, Ottawa, 1973.

Loudon, J.C.
An Encyclopedia of Cottage, Farm and Villa Architecture and Furniture. Longman, Brown, Green and Longman's, London, 1846.

Lutman, John H.
The Historic Heart of London. Corporation of the City of London, London, 1977.

Maitland, Leslie
"The Queen Anne Revival Style in Canadian Architecture." Manuscript in process, Environment Canada, Parks Service, Ottawa, 1986.

Marsan, Jean-Claude
Montréal en évolution. Fides, Montreal, 1974.

Marshall, Philip C.
"Polychromatic Roofing Slate of Vermont and New York." *APT Bulletin.* Vol. 2, No. 3 (1979), p. 77.

Maurault, Olivier
La paroisse; histoire de l'église Notre-Dame de Montréal. Thérien Frères, Montreal, 1957.

Mayrand, Pierre
"Les matériaux de couverture en Nouvelle-France aux XVIIe et aux XVIIIe siècles." *APT Bulletin*, Vol. 2, Nos. 1-2 (1970), pp. 71-73.

McKee, Harley J.
"Slate Roofing." *APT Bulletin*, Vol. 2, Nos. 1-2 (1970), pp. 77-84.

Merriman, Mansfield
"The Slate Regions of Pennsylvania." *Stone*, Vol. 17, No. 2 (July 1898), pp. 77-90.

Metal Worker, Plumber and Steam Fitter (New York)
"Who does the Slate Roofing in Your Town?" Vol. 90 (April, 1918), pp. 522-23.

Moogk, Peter
Building a House in New France. McClelland and Stewart, Toronto, 1977.

Morris, William
"The Influence of Building Materials Upon Architecture." Delivered before the Art Worker's Guild at Barnard's Inn Hall, London, January 1892. In *The Collected Works of William Morris.* Longman's, Green and Company, London, 1914. 24 vols. Vol. 22, pp. 391-409.

Nader, George A.
Cities of Canada. Macmillan, Canada, 1975. 2 vols. Vol. 1.

National Slate Association
Slate Roofs. Reprint of 1926 edition. Vermont Structural Slate Company, Fair Haven, Vermont, 1977.

Nicholson, Peter
Encyclopedia of Architecture. Johnson, Fry & Co., New York, 1858. Vol. 1.

Noppen, Luc
Les églises du Québec (1600-1850). Éditeur officiel du Québec/Fides, Quebec, 1977.
———. *Notre-Dame de Québec: son architecture et son rayonnement (1647-1922)*. Éditions du Pélican, Quebec, 1974.

Noppen, Luc, Claude Paulette, and Michel Tremblay
Québec: trois siècles d'architecture. Libre Expression, Québec, 1979.

Old-House Journal, The (Brooklyn, New York)
"Slate Roofs." Vol. 8, No. 5 (May 1980), pp. 49-55.
———. "Roofing Materials Compared." Vol. 11, No. 3 (April 1983), p. 57.
———. "Substitute Roofings — Credible Stand-Ins for Clay Tile, Slate, and Wood." Vol. 11, No. 3 (April 1983), p. 61-63.
———. "Replacing a Slate." Vol. 11, No. 3 (April 1983), p. 66.
———. "Restoration Product News — Slate Shingles." Vol. 11, No. 3 (April 1983), p. 73.

Parks, W.A.
Report on the Building and Ornamental Stones of Canada. Canada, Department of Mines, Mines Branch Publication No. 279. King's Printer, Ottawa, 1914. 5 vols. Vol. 3.

Quebec. Department of Colonization, Mines and Fisheries.
Report on Mining Operations in the Province of Quebec, 1921. King's Printer, Quebec, 1922.

Ritchie, Thomas
Canada Builds: 1867-1967. National Research Council of Canada, Ottawa, 1967.

Ruskin, John
The Stones of Venice ... John Wiley, New York, 1851. 3 vols. Vol. 1.

Scott, M.H. Baillie
"The Art of Building." *Construction*. Vol. 3, No. 4 (April 1910), pp. 97-98.

Segger, Martin, and Douglas Franklin
Victoria: A Primer of Regional History in Architecture, 1843-1929. A Pilgrim Guide to Historic Architecture, Watkins Glen, N.Y., 1979.

Simcoe, Elizabeth
Mrs. Simcoe's Diary. Ed. Mary Q. Innis. Macmillan, Toronto, 1965.

Sloan, Samuel
Sloan's Homestead Architecture ... J.B. Lippincott and Company, Philadelphia, 1861.

Stevenson, J.J.
House Architecture. Macmillan and Co., London, 1880 Vol. 2.

Stringer, Charles H.
"Modern Roofing Materials — Part III: Asbestos Shingles and Slates." *The American Architect.* Vol. 114, No. 2231 (25 Sept. 1918), pp. 62-65.

Taché, J.C.
Descriptive Catalogue of the Productions of Canada Exhibited in Paris in 1855. G.A. Pinard-Dentan and Co., Paris, 1855.

Thomas, Christopher A.
"Dominion Architecture: Fuller's Canadian Post Offices, 1881-96." Master's thesis, University of Toronto, Toronto, 1978.

Traquair, Ramsay
The Old Architecture of Quebec. Macmillan, Toronto, 1947.

Turner, Philip J.
"Houses at Montreal, Quebec." *Construction*, Vol. 18, No. 6 (June 1915), pp. 265-73.
———. "Recent Houses in Montreal and Westmount. *Construction*, Vol. 9, No. 6 (June 1916), pp. 198-200.

United States. Geological Survey.
Mineral Resources of the United States. Government Printing Office, Washington, 1904-30.

Vaux, Calvert
Villas and Cottages ... Harper and Brothers, New York, 1857.

Vincent, Elizabeth
"The Royal Engineer's Use of Roofing Materials in British North America 1820-1870." Manuscript in progress. Environment Canada, Parks Service, Ottawa, 1987.

Viollet-le-Duc, E.
Dictionnaire raisonné de l'architecture française du XIe au XVIe siècle. V.A. Morel & Cie., Paris, 1875. 10 vols. Vol. 1.

Waite, Diana S.
"Roofing for Early America." In *Building Early America: Contributions Toward the History of a Great Industry.* Ed. Charles E. Peterson. Chilton Book Company, Radnor, Pennsylvania, 1976.

Weld, Isaac, Jr.
Travels through the States of North America and the Provinces of Upper and Lower Canada during the years 1795, 1796 and 1797. N.P., London, 1800. Vol. 1.

Whedon, Milford D.
"The New York-Vermont Slate Belt." *Stone*, Vol. 28 (1909), pp. 214-18.